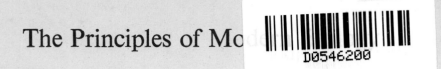

The Principles of Mo

Statistics for Biology

A practical guide for the
experimental biologist

Second edition

O. N. BISHOP

LONGMAN

LONGMAN GROUP LIMITED
London
*Associated companies branches and representatives
throughout the world*

© O. N. Bishop 1966
© *Second edition* Longman Group Ltd 1971

First published 1966
Third impression with SI units 1969
Second edition 1971
Third impression 1975
ISBN 0 582 32314 2

*Printed in Hong Kong by
Sheck Wah Tong Printing Press*

Foreword

Within the last two or three decades Biology has made a most impressive growth spurt, and biologists have moved forward to a new understanding of some of the major problems of life. The tools that have made these steps possible have largely been made available by contemporary advances in the disciplines of physics, chemistry, and mathematics, and their application in the precise skills of engineering. It has been possible to establish a closer unity between the several biological sub-disciplines, and the integration achieved has enabled substantial progress to be made.

Biology, like all the sciences, is dependent upon observation and measurement; indeed, it has not very long emerged from a phase dominated by qualitative observation, into a phase in which observations are largely in the form of quantitative measurements. Biological material is inherently variable, and it often becomes necessary to express data in such a way that this variation may not be misleading.

The present book must be seen, therefore, as complementary to the other books of the series in providing a practical means for assessing the quantitative results of practical work. One can hardly overemphasise the need for turning, wherever possible, from text book to organism, both in field and laboratory; it is intended that this book should be of help in arraying and testing the validity of some of the data so obtained. In the more recent phase of measurement, statistics has become a vitally necessary weapon in the armoury of the biologists.

D. A. COULT

University of Liverpool

Contents

Preface to the first edition

Statistical analysis is becoming increasingly important as a tool in biological investigation, not only in the university and in industry, but also in schools. Most biologists are not so much concerned with acquiring detailed knowledge of the mathematical theory of statistics, as in learning how to select and how to use those statistical techniques which will help them interpret the results of their experiments. Accordingly, I have emphasised the practical applications of statistical analysis, and have kept mathematics to a minimum. There is a limit to the number of techniques which can be dealt with in an introductory text of this size. Sooner or later the experimenter will have some results which cannot be properly handled by the methods described here. Then it will be time to consult either the more advanced texts, or a qualified statistician. My hope is that the reader will gain sufficient from this book to know when the methods described here are inadequate, and to be able to pass easily to the advanced texts or to converse intelligibly with the expert.

I wish to acknowledge the invaluable help and advice given by Mrs Patricia Cooke, B.Sc., of the United Sheffield Hospitals Centre for Human Genetics. During the preparation of the manuscript she has kept a watchful eye for errors of principle, of fact, and of computation. I thank her for this, as well as for her suggestions for ways of improving the content of the text. Nevertheless, responsibility for mistakes of any kind remains my own. I thank my wife, Mrs Audrey Bishop, B.Sc., and Dr M. D. Casey, both of the United Sheffield Hospitals Centre for Human Genetics, Dr H. W. Woolhouse and Dr D. J. Anderson, both of the Department of Botany, The University, Sheffield, and W. J. Varley, of Worksop College, for their helpful advice and for their assistance in providing data for some of the worked examples.

I am indebted to the late Sir Ronald Fisher, F.R.S., and to Dr Frank Yates, F.R.S., of Rothamsted, also to Messrs Oliver and Boyd Ltd, Edinburgh, for permission to reprint Tables III, IV, V, and VII from their book *Statistical Tables for Biological, Agricultural and Medical Research*. I am indebted also to Addo Ltd, and to the Muldivo Calculating Machine Co Ltd, for their help and advice.

Worksop College   O. N. B.

Preface to the second edition

The first edition has been widely used by students and teachers in university biological departments, in technical colleges, and in the sixth-forms of secondary schools, both in the United Kingdom and overseas. It has also been of practical help to a number of research workers in various fields of biology and medicine. In revising it for the second edition I have tried to extend its usefulness by enlarging the chapter on the planning of experiments, and by including a completely new chapter, which deals with distribution-free tests. The distribution-free tests will be particularly useful to biologists, and some extra tables have been included in this edition for use with these tests.

Wiseton   O. N. B.

The plan of this book

The first five chapters describe the main terms and concepts.

The next six chapters deal with more complex techniques, and may be omitted on first reading.

The last three chapters and the Statistical Tables which follow them are not for reading but for reference.

CHAPTER 12 is a key to selecting the most appropriate statistical tests,

CHAPTER 13 gives concise practical instructions for performing these tests, and

CHAPTER 14 explains how to use a hand calculating machine to carry out the tests more speedily.

1 Why use statistics?

When the results of an experiment or of a series of observations have been recorded, there arises the problem of interpreting them and deriving reasonable conclusions. Sometimes the results are self-evident and interpretation is easy. For example, if twenty laboratory rats are first infected with a disease organism, and then half of the rats are given a dose of a new drug, we would be justified in thinking that the drug was effective if all of the dosed rats survived while all of the others died of the disease soon afterwards. Results like these do not need rigorous statistical analysis, but unfortunately the results of most biological experiments are seldom so decisive.

Table 1

Dry weights of sunflower plants grown in water culture

		Dry weight (mg)			
Batch	Culture solution	Plant 1	Plant 2	Plant 3	Average
A	Complete	1172	750	784	902
B	Lacking magnesium	67	95	59	74
C	Lacking nitrogen	148	234	92	158
D	Lacking micro-nutrients	297	243	263	268

Table 1 shows the results of an experiment carried out by a worker who was interested in the role of nitrogen, magnesium, and certain micro-nutrients in the metabolism of sunflowers. He grew twelve sunflower plants in water culture for nine weeks, under uniform conditions of lighting and temperature. Each plant was in a separate pot and was supplied, throughout the experiment, with one type of culture solution. Thus all plants were treated in a uniform fashion except in respect of a single factor—the type of culture solution supplied. The three plants in Batch A were grown in a complete culture solution. Those in Batch B received a solution which differed from the complete solution only in that it lacked magnesium. Similarly, those in Batch C received a solution lacking only nitrogen, and those in Batch D received a solution lacking only the micro-nutrients (boron, manganese, zinc, and copper). At the end of the experiment each plant was taken separately, chopped into pieces, dried in an oven, and weighed, giving the results shown in the table.

1

Assuming that dry weight is a valid measure of growth, at once it can be seen that growth was greatest in the complete solution (Batch A). Plants in the complete solution became heavier than any of the plants grown in the other solutions. There need be no hesitation in arriving at this conclusion, for though the plants of Batch A varied rather widely in weight (from 750 mg to 1 172 mg), the lightest of them was over 2·5 times as heavy as the heaviest of the other plants.

Next consider whether plants lacking micro-nutrients (Batch D) grew more than those lacking nitrogen (Batch C). This is not so easily decided by mere inspection of the table. Although the average weight of Batch C is less than that of Batch D, there is considerable variation within each batch, and the heaviest of Batch C is almost as heavy as the lightest of Batch D. It is possible that these figures are the result of the worker having unwittingly selected three poor specimens for growing in the nitrogen-deficient solution, and three better specimens for the other solution, in spite of having taken all possible measures at the beginning of the experiment to make sure that the twelve plants were uniform in all ways. Thus although it appears likely that there is a real difference in growth between the two batches, their different average weights could be explained in another way.

In comparing the weights of Batch C with the weights of Batch B it is even more difficult to reach a definite and reliable decision. Both sets of figures show wide variation, and there is an overlap between them. To form a definite interpretation something more precise than personal opinion is required; only careful statistical analysis can supply the answer. The analysis will not necessarily give an affirmative statement, such as: 'Plants lacking nitrogen in general grow less than those lacking micro-nutrients.' This degree of certainty is not to be expected, since the dry weights themselves are not separated into two absolutely distinct groups, and there will be a corresponding element of doubt in the statement made after the statistical analysis. A more realistic statement would be: 'The probability that plants lacking nitrogen grow more, on average, than plants lacking micro-nutrients is approximately 19 to 1.' The statement is *qualified* by a numerical estimate of the probability of its truth. This is the function of statistical analysis—it gives a *measure* of the *probable* truth of statements. Without analysis, the most that can be done is to qualify statements with imprecise, subjective, and sometimes misleading phrases such as: 'It is likely that . . .', 'It is possible that . . .', or 'It seems that . . .'.

2

Biological experiments deal with living things, and it is characteristic that no two individuals are exactly alike. This variation between individuals is the basis of the evolution of new forms of life. Yet, in spite of variability it is important to be able to make statements which may be applied to whole groups of individuals, even to all the individuals of a species. In the example given above it was evident that no two sunflower plants are exactly alike, even when grown under identical conditions, yet the worker was attempting to make general statements about the effects of different culture solutions upon sunflower growth. When the effects are large, the conclusions are self-evident, but when there are effects which are only slightly greater than the natural differences between individual plants, the objectiveness and precision of statistical analysis is required. This is the only valid and reliable basis for the interpretation of the results. It is essential that the final conclusions shall be firmly based.

2 Some terms

During an investigation of the composition of food materials, the water content of five different turnips was measured, and this set of results was obtained:

Turnip No.	1	2	3	4	5
Water content (as a percentage of fresh weight)	90·7	90·8	92·3	94·0	96·1

In this series of observations the percentage of water is a **variable**, a characteristic of each turnip which can differ in value from one turnip to another. The variable is the raw material upon which we work when carrying out an analysis. Such a variable as the one given above is a **continuous variable,** because it may take *any* value within a certain range. In the example the variable ranges from 90·7% to 96·1%, and one may reasonably suppose that with other turnips one would obtain percentages of any value within this range, and slightly above it and below it as well. Variables such as height, weight, and the percentages of this example are continuous because of the nature of the phenomena they represent. Living organisms grow continuously, and absorb or lose water continuously, not in discrete steps.

In other biological investigations we may take measurements of quantities of a different kind. These are called **discrete** or **discontinuous variables.** For example, there may be 1, 2, 3, 4, 5, or 6 peas in a pod, but there are never 4·35 peas. Fractions of peas do not exist as peas, and their numbers must be integers. Thus the number of peas is discontinuous. The number of ray-florets on a daisy, the number of tentacles on a hydra, and the number of animals caught in a trap are all examples of discontinuous variables.

The results quoted above are measurements of the variable for five turnips, the largest number of turnips assayed in the time available. Ideally one would have measured the percentage of water in all turnips which exist. Clearly this is impossible, but, if this could have been done, one could have then made a statement about water content which would have been true for all turnips. This introduces

4

the concept of **population.** In this case the population is 'all the turnips which exist'. Similarly, all the antirrhinums in a garden might be considered to be a population; so might all the snails in a pond. The interpretation of the term, population, depends upon its context. For some purposes, one may regard a garden as containing a whole population of antirrhinums; the population is defined as 'all the antirrhinums in the garden'. With this population one can experiment upon the whole, or upon a few plants, collected as a **sample.** On another occasion one may regard the whole garden-full of antirrhinums as a sample from a larger population. The population might then be defined as 'all the antirrhinums growing in gardens in England'.

The discussion of the population concept has introduced the idea of a sample. One collects a sample of convenient size (a convenient number of turnips was five) in order to perform tests or take measurements. This is done to provide information about the population from which the sample was drawn. One assumes that the sample is truly representative of the population, and the next step is to make *estimates* about the population, basing the estimates upon what is *known* about the sample. This procedure is one of the main operations of statistical analysis, and the next example will illustrate one of the ways in which it is done.

It was decided to try to estimate the reproductive potential of antirrhinum plants, taking the average number of seeds produced by each plant as a measure of this. The simplest plan was to find the average number of flowers on each plant, and the average number of seeds formed by each flower, and then calculate the product of these two quantities. The plants examined were of the cultivated species, *Antirrhinum majus*, of which there were 43 plants of a red-flowered variety (Sutton's *Eclipse*) growing under reasonably uniform conditions in a garden.

In mid-September the flowers, flower-buds, and ripening capsules were counted for each plant, but to simplify the calculation and this description the flower-buds and ripening capsules will be included under the single term, flower. The variable is therefore the number of flowers on each plant—a discontinuous variable.

One might consider that the whole population of plants had been examined, since all the plants in the garden had been included. In one sense this was true; the population could have been defined as all the plants growing in the garden at that time. In another sense it was only a sample. With more packets of seed of the same variety,

5

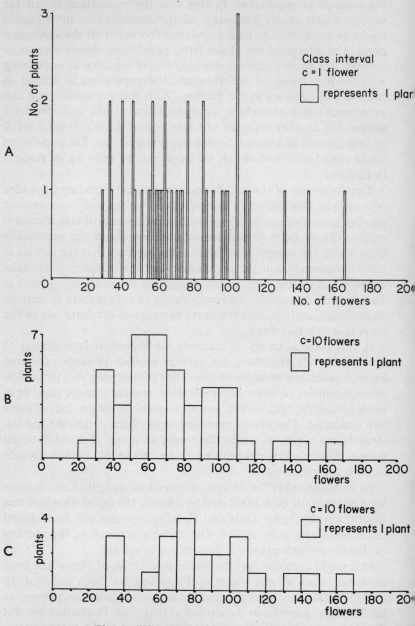

Fig. 1. Histograms illustrating the frequency distributio

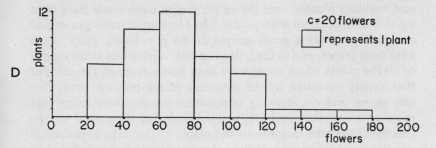

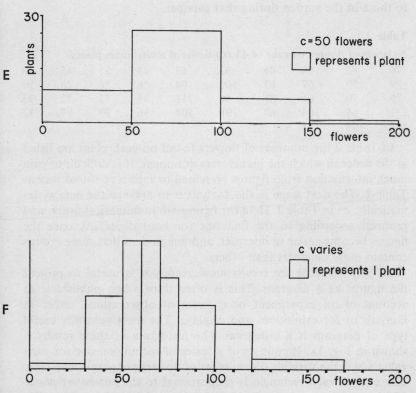

of the number of flowers on antirrhinum plants.

and seedlings planted over the entire garden there could have been a population of about 4000 plants. The 43 plants actually grown and counted were only a small sample of the population which could have been grown, and in turn, this was only a relatively small sample of all the plants which could have been grown from all the seeds of that variety produced by the seedsman in the previous year. One can go on and on, defining populations progressively larger and more encompassing, and less easily imaginable. One need not be concerned that the majority of the plants of the large hypothetical populations do not exist. It is in this hypothetical population—all red-flowered antirrhinum plants—that the experimenter's interest lay. He did not wish to restrict his findings to the few in the garden, but by examining the sample of 43 plants, to estimate the characteristics of the hypothetical population, the population of antirrhinum plants of the *Eclipse* variety when grown under conditions similar to those in the garden during that summer.

Table 2

Numbers of flowers on each of 43 red-flowered antirrhinum plants

76	62	54	46	33	63	45	55	45	39
39	59	67	63	109	98	58	56	28	50
76	70	61	85	56	111	85	72	32	73
131	64	104	60	91	104	96	99	87	32
104	144	165							

In Table 2 the numbers of flowers found on each plant are listed in the order in which the plants were examined. It is difficult to gain much information from figures presented in such a confused way as Table 2. The next stage in the analysis is to arrange the data systematically, as in Table 3. Here the figures are in numerical order, and grouped according to the first one (or two) digits. At once the figures become easier to interpret, and one notices that some groups contain more members than others.

To help visualise the results more readily it is useful to present the figures as a diagram. This is often done when publishing an account of an experiment or a series of observations, either in journals or for exhibitions and displays. The most generally useful type of diagram is a **histogram**. The histogram of these results is shown in Fig. 1A. It consists of a series of rectangles, one for each value which the variable (i.e. number of flowers per plant) may take. The *area* of each rectangle is proportional to the *number of plants* which are found to have the given number of flowers on them. We

Table 3

Data of Table 2, arranged systematically

Number of flowers on each plant

28						
32	32	33	39	39		
45	45	46				
50	54	55	56	56	58	59
60	61	62	63	63	64	67
70	72	73	76	76		
85	85	87				
91	96	98	99			
104	104	104	109			
111						
131						
144						
165						

call this number of plants the **frequency** with which this given number of flowers occurs. For example there are 2 plants with 56 flowers, so the frequency of the variable value '56 flowers' is 2; similarly the frequency of 87 flowers is 1, and the frequency of 104 flowers is 3.

The area of each rectangle is proportional to the frequency, but in this diagram the rectangles are all of the same width, so that the heights too are proportional to frequency. For example, there is one plant with 28 flowers, so this is represented by a rectangle of unit area and height. There are two plants with 33 flowers each, so these are represented by a rectangle of double the area, and two units high. The scale on the left of the histogram indicates units of height and from this one can read the frequency of each value of the variable. You might be wondering why there has been so much emphasis on area, when height is simpler to draw and use, but later there will be instances where height is of no use and one must revert to considering areas.

The histogram in Fig. 1A is too irregular to give much information, but to obtain a clearer picture the plants can be grouped into **classes**. For instance, all those plants with between 30 and 39 flowers go into one class, all those with between 40 and 49 plants go in another class, and so on. The result of classing the data in this way is shown in Table 4, and the corresponding histogram is Fig. 1B. This arrangement into classes had already been done when setting out Table 3; all that remained was to count the number of individual plants on each line of the table. The new histogram has fewer gaps and is more regular in outline. It becomes possible to pick out classes which have higher frequencies than others, such as the two classes 50–59 and 60–69. Some order is beginning to emerge from the confused set of figures of Table 2.

Table 4

Data of Table 2, arranged in classes of 10

Class (No. of flowers per plant)	Frequency
0– 9	0
10– 19	0
20– 29	1
30– 39	5
40– 49	3
50– 59	7
60– 69	7
70– 79	5
80– 89	3
90– 99	4
100–109	4
110–119	1
120–129	0
130–139	1
140–149	1
150–159	0
160–169	1

Total = 43 plants

Even when the plants are put into classes, there are still some irregularities in the histogram. This is mainly because each class contains only a few members, and one plant more or less makes a lot of difference to the appearance of the histogram. Suppose that the two plants with 39 flowers had developed just one more flower each. They would then have to be transferred to the 40–49 class, making the class frequencies 0, 0, 1, 3, 5, 7, 7, ... This is a much more steady rise from zero to the peak of the histogram. However, one cannot add flowers to the plants to suit the appearance of the histogram—one must make do with the results one already has, or try to improve them in some other way. There are two ways of doing this:

1. *By assembling more data:* With a greater number of plants, there will be more in each class and irregularities will be relatively less

10

noticeable. The effect of this cannot be demonstrated in this example, because no more plants are available. Instead, see what happens with *fewer* figures. Compare Fig. 1c (which has been prepared by using the figures from only the last 22 plants to be counted) with Fig. 1B (which includes all 43 plants). Increasing the number of plants from 22 to 43 has made some reduction in the irregularity of the histogram. Increasing the number of plants to 100 or more would probably give an even more regular histogram. Unfortunately there were no more plants of this kind in the garden, and so the second approach to the problem must be tried. Before passing to this, it is worth considering that perhaps there is *no* regularity underlying the figures. Perhaps attempts to obtain a more regularly shaped histogram will be fruitless.

2. *By using fewer classes:* Each class will then contain more individuals (achieving the same effect as in the method above). In Table 4 the data are grouped into classes of ten; each class begins with a number of flowers which is ten more than that of the beginning of the class below it; one says that the **class interval** (C in Figs. 1 and 3)

Table 5

Data of Table 2, arranged in classes of 20

Class (No. of flowers per plant)	Frequency
0– 19	0
20– 39	6
40– 59	10
60– 79	12
80– 99	7
100–119	5
120–139	1
140–159	1
160–179	1
	Total = 43 plants

is ten flowers. Table 5 shows what happens when the class interval is increased to 20 flowers. The histogram (Fig. 1D) now shows a steady rise to a peak in the 60–79 class, and falls away steadily as the higher classes are reached. This is probably the most useful histogram which can be obtained from this data.

If the class interval is increased farther, one obtains histograms which are less informative, because there are too few classes to define the shape of the histogram. Figure 1E has a class interval of 50 flowers, and detail is lost because there are only four columns to indicate the distribution of the frequencies.

To sum up, the class interval is chosen so that the classes are large enough to reduce the effects of chance variations in frequency, but are not so large as to make the outline of the histogram too crudely defined. This is not to say that all irregularities *must* be smoothed out Occasionally a histogram will have two or even more peaks, which are not due to the chances of sampling. For example, this may happen when a population contains two distinct types or varieties which have been sampled together. The histogram will have two peaks, indicating a *bimodal* distribution.

In all the histograms drawn so far, the columns have been of equal width because the class intervals have been of equal size. Though this is desirable for the type of calculation to be performed later, it is not necessary if one is plotting a histogram simply to illustrate the data. Figure 1F shows a histogram with unequal classes, based upon the class intervals listed in Table 6. A smaller class interval has been

Table 6

Data of Table 2, arranged in unequal classes

Class (No. of flowers per plant)	Frequency	Class interval	Factor (= 10/C.I.)	Relative height (= frequency x factor
0– 29	1	30	$\frac{1}{3}$	$\frac{1}{3}$
30– 49	8	20	$\frac{1}{2}$	4
50– 59	7	10	1	7
60– 69	7	10	1	7
70– 79	5	10	1	5
80– 99	7	20	$\frac{1}{2}$	$3\frac{1}{2}$
100–119	5	20	$\frac{1}{2}$	$2\frac{1}{2}$
120–170	3	50	$\frac{1}{5}$	$\frac{3}{5}$

Total = 43 plants

used around the centre of the distribution, where frequencies are highest, while at the extremes the class intervals are larger. Thus it is possible to compromise differently between frequency and class interval at different parts of the histogram. This figure is drawn to the same scale as the others, as shown by the *area* scale at the top left of each histogram. In this histogram there are no markings on the vertical scale, for these would be merely misleading as it is not now possible to read off frequency from the height of each column. In plotting the histograms the relative heights of the columns were calculated according to the method shown in Table 6. To keep area proportional to frequency the heights of the wider columns (large class intervals) are relatively reduced in height, according to the factors calculated in the table.

You may be wondering why it is necessary to go to the extra trouble of calculating areas for plotting histograms like Fig. 1F, when it would be much easier to work with column heights as before. As stated on p. 9, it is the underlying areas which must determine the heights of columns of the histogram, and it is only when all class intervals are equal that heights too are proportional to frequency. Later (p. 29) it will be explained how a histogram can be made to approach the shape of a special mathematical curve which is of great theoretical importance. It is only if the histogram is drawn with area proportional to frequency that the histogram and the curve can be properly compared.

When drawing each of the histograms (Figs. 1A to 1F) the data have been grouped in classes and the frequency with which observations fall into each class have been recorded. In this way the **frequency distribution** of the variable has been obtained. The histograms help one to visualise the distributions. To proceed further one must do more than simply look at histograms. The frequency distributions must be described in mathematical terms, so that it becomes possible to make quantitative comparisons between one distribution and another. This forms the subject of the next chapter.

PROBLEMS

Here are several sets of data, presented just as they were collected. Arrange each set systematically to find the frequency with which each value of the variable occurs. Then group the data into classes and draw histograms to illustrate the frequency distributions. Work on these problems will be continued later in the book, so it is best to use a loose-leaf binder. The answers are on p. 206.

1 These figures were obtained by counting the number of ray-florets on 51 capitula of daisy (*Bellis perennis*):

```
55 47 42 44 51 55 48 48 48 41 40 51 49 40 50 39 42 50 48 52
44 43 43 41 49 36 40 46 44 48 49 42 45 47 38 42 41 44 52 50
44 44 42 37 41 34 42 43 41 32 37
```

Tabulate the figures systematically to show the frequency distribution of numbers of ray-florets. For the histogram, group them into the classes 30–34, 35–39, 40–44, 45–49, 50–54, and 55–59 ray-florets. What value has the class interval?

13

2 In an experiment designed to test the effect of mineral fertiliser supply upon the growth of radishes, the control batch of radish seeds was germinated in vermiculite to which no fertiliser had been added. The experimental batch was germinated in vermiculite to which National Growmore fertiliser had been added. To detect an effect at an early stage the width of a cotyledon was measured on each seedling, when the plumules were only just beginning to show. The cotyledon widths, in mm, were:

No fertiliser	18	19	12	16	14	17	13	16	15	14	20	17
	10	8	17	28	19	12	16	16	19	17	12	
Fertiliser provided	29	17	21	21	30	19	17	17	19	11	18	
	20	19	19	27	14	25	19	20	26	9	11	24	24	27	15
	24	25	16	19	16	23	18	23				

Draw a histogram for each treatment, using classes 5–9, 10–14, 15–19 mm, and so on. From inspection of the histograms, what can be said about the effect of fertiliser treatment at this stage in the life of the plant?

3 A batch of locust nymphs was weighed, with these results, in grams:

0·95 1·18 1·05 1·22 0·82 0·93 0·68 0·52 0·70 0·67 0·57 0·59 0·35 0·92 0·67 0·54 1·12 0·89 0·84

Try plotting several histograms, using different class intervals.

4 Fifty-one mature flowers of creeping buttercup (*Ranunculus repens*) were dissected and the numbers of stamens and carpels in each flower was determined. The figures below are given in pairs, the numbers of stamens and carpels in the same flower. The pairing will be needed in a later problem, but does not affect the working here.

Stamens 60 54 55 70 64 72 53 40 43 63 65 53 57 58 51 67 50 42 43
Carpels 30 21 31 32 27 29 19 21 22 20 19 30 25 31 28 31 35 34 36

Stamens 39 46 38 49 46 47 48 40 49 37 35 47 58 74 64 61 77 62 57
Carpels 26 32 23 32 36 29 28 29 30 26 26 28 24 22 33 36 31 28 25

Stamens 75 73 78 52 45 50 62 55 58 57 55 49 40
Carpels 34 30 36 23 33 26 39 32 30 33 44 35 42

Plot histograms to illustrate the frequency distributions of (1) numbers of stamens and (2) of carpels, using a class interval of 10. From inspection of the histograms what can be said about the relative numbers of carpels and stamens in the flowers of *Ranunculus repens*?

5 Some individuals of a species of Ostracod crustacean were placed in a drop of pond water on a microscope slide; the cover-slip was gently lowered to prevent the animals from swimming. Those trapped near the edge of the cover-slip were examined under the microscope, and the respiratory movements of the abdominal appendages were timed for each of 10 individuals. The experiment was first performed outdoors, at a temperature of 13°C, and then repeated with another 10 ostracods indoors, at 20°C. The respiratory movements, in beats per 15 seconds were:

| *At 13°C* | 34 | 37 | 33 | 35 | 33 | 42 | 33 | 32 | 32 | 39 |
| *At 20°C* | 45 | 44 | 60 | 69 | 59 | 45 | 48 | 55 | 50 | 53 |

Using classes 30–34, 35–39, 40–44, and so on, plot two histograms. What can be said about the effect of temperature upon the rate of respiratory movement?

6 Certain plants of *Ranunculus acris* growing beside a hedgerow appeared to the eye to have markedly smaller flowers than is usually found in this species. To obtain quantitative confirmation of this visual impression, measurements were made of flower diameter on the 'small-flowered' plants and on adjacent plants which had apparently normal flowers. The diameters, in mm, were:

Small-flowered 15 9 13 8 13 9 7 10 11 9 10 11 11 12 12 11
Normal-flowered 15 19 14 20 17 12 21 17 14 16 12 19 12 17 19 19 19
       19 17 19 19 19 20 25 18 20 21 22 16 19 19 23 21 20
     ˇ 19 18 18 15 20 15 19 20 18 21 25 19

Part of the variation in diameter is due to the fact that some of each type of flower were still opening from the bud. Each type included a majority of mature flowers at the time of measurement, so there is no possibility that the small-flowered type were simply later opening. Draw histograms to show the frequency distribution of flower diameter in the two types, choosing a suitable class interval. Do the histograms confirm the visual impression of a difference in size?

3 Describing a distribution in mathematical terms

Most distributions met with in biology are like the one illustrated in Fig. 1D (from Table 5). There is a peak in the middle, and the distribution tails off more or less symmetrically at higher and lower values of the variable. The first feature to describe mathematically is the position of the peak. There are three ways of doing this:

Mode This is defined as the most frequent class. In this example, the modal class is 60–79 flowers. This is a simply determined quantity, but it can sometimes be misleading as a guide to the nature of the distribution. For instance, in Fig. 1A, with ungrouped data, the mode is 104 flowers, which is very near the upper end of the distribution.

Median This is the middle observation, when they are arranged in order of magnitude, as in Table 3. There are 43 observations, so the middle one is the 22nd, with 21 observations smaller than this and 21 observations greater. Counting along the rows, the 22nd observation has the value of 64 flowers; so the median of this distribution is 64 flowers. The median falls within the modal class of Table 5, though this need not always be so.

Since the areas of the columns of a histogram are proportional to the number of observations (frequency), it follows that a vertical line drawn through the median will cut the area of the histogram into two equal halves. The area to the right of the line (observations greater than the median) will be equal to the area to the left of the line (observations less than the median).

Mean This is the arithmetic mean, or average, of the observations. Mathematically it is defined as follows:

If the observations are represented by $x_1, x_2, \ldots x_n$, their mean, $\bar{x}$ (say: 'x bar'), is calculated by summing the observations and dividing the sum by the number of observations, which is n.

$$\bar{x} = \frac{1}{n}(x_1 + x_2 + \ldots + x_n)$$

16

There is a convenient shorthand way of indicating when a series of quantities like the observations, $x_1, x_2, \ldots, x_n$, are to be added together. Instead of writing:

$$x_1 + x_2 + \ldots + x_n$$

we write: $\qquad\qquad\qquad \Sigma x \qquad\qquad$ (say: 'sum of x')

Similarly, Σx^2 means:

$$x_1^2 + x_2^2 + \ldots + x_n^2$$

—the **sum of squares** of all observations.

This is *not* the same as $(\Sigma x)^2$, which means:

$$(x_1 + x_2 + \ldots + x_n)^2$$

—the *square of the sum* of all observations.

Using this shorthand convention, the equation for the mean may be rewritten:

$$\bar{x} = \frac{\Sigma x}{n}$$

In the example (Table 3) the mean is calculated like this:

$$\cdot\, \Sigma x = 28 + 32 + 32 + \ldots + 144 + 165 = 3147 \text{ flowers}$$
$$n = 43 \text{ plants}$$

$$\therefore \quad \bar{x} = \frac{\Sigma x}{n} = \frac{3147}{43} = 73 \cdot 19 \text{ flowers per plant}$$

The mean differs from the median of this distribution, though like the median it is within the modal class. However, in other distributions the mean need not be equal to either the median or the mode. These three ways of describing the position of the peak of the distribution are independent of one another. Each has its uses in differing circumstances, but the one used most often is the mean.

Another formula for the mean

Table 3 contains 43 observations, but there are only 34 different values of the variable, because for certain values there are two or three plants bearing the same number of flowers. Some of the values of the variable are more frequent than others and the data of Table 3 can be set out in the form of a frequency distribution with class interval of one. This has been done in Table 7, but to save space, only the lower values (the first three rows of Table 3) have been included. In the first

Table 7

Data from the first three rows of Table 3, arranged as a frequency distribution (class interval, $c = 1$)

Class (No. of flowers per plant) x	Frequency (No. of plants with x flowers) f	fx
28	1	28
29	0	0
30	0	0
31	0	0
32	2	64
33	1	33
34	0	0
35	0	0
36	0	0
37	0	0
38	0	0
39	2	78
40	0	0
41	0	0
42	0	0
43	0	0
44	0	0
45	2	90
46	1	46
	$\Sigma f = 9$	$\Sigma fx = 339$

column are all the values that the variable can take; this is the term x in the definition of the mean. In the second column is the frequency, f, with which each value of the variable occurs. The figures in the third column are obtained by multiplying each value of x by its frequency, f. This product is indicated by the term fx at the head of the third column. The total of this column is Σfx, which means $f_1 x_1 + f_2 x_2 + \ldots + f_n x_n$. This is the sum of all the products. The term fx is a convenient device for use in calculating the mean. Instead of adding the values separately:

$$28 + 32 + 32 + 33 + 39 + 39 + 45 + 45 + 46 = 339$$

the same result is obtained by multiplying each value by its frequency and summing the products:

$$(1 \times 28) + (2 \times 32) + (1 \times 33) + (2 \times 39) + (2 \times 45) + (1 \times 46)$$
$$= \quad 28 \quad + \quad 64 \quad + \quad 33 \quad + \quad 78 \quad + \quad 90 \quad + \quad 46 \quad = 339$$

With low frequencies, as in this example, this method of calcula-

18

tion is not much quicker, but it saves much time and effort when frequencies are high.

The term Σf is the total number of observations recorded in the table. Algebraically, $\Sigma f = n$, and we may write the equation for the mean in terms slightly different to those used on p. 17.

$$\bar{x} = \frac{\Sigma fx}{\Sigma f}$$

This alternative equation does not signify a different sort of mean, it merely indicates a different way of performing the calculation. In either case the answer will be the same.

PROBLEMS

1–6 The working of these examples can now be taken a stage further. Using the histograms, pick out the modal class of each distribution. Using the original unclassed data, arranged systematically, find the median and the mean of each distribution. Instructions for methods of calculating means are given in brief form on pp. 154–7. The second method described there is a variant of the method given on p. 18, and uses a 'working mean' to lessen the amount of arithmetic required. It also includes a check on the correctness of calculation. It is worth while to practise this method when working the larger sets of data, especially Problems 2, 4, and 6. The technique for calculating means by machine is given on p. 179.

7 To estimate the numbers of daisy (*Bellis perennis*) and ribwort (*Plantago lanceolata*) plants growing on a lawn, a circular frame was thrown down at random 100 times, and the numbers of plants of the two species so ringed were counted. These two distributions were obtained:

No. of plants ringed	0	1	2	3	4	5	6	7	Total throws
Frequency Daisy	47	20	12	7	9	1	1	3	100
(no. of throws) Ribwort	63	20	11	4	1	1	0	0	100

The table shows, for example, that the ring encircled four daisy plants in nine throws out of the 100; on 47 throws, no daisy plants were ringed. Plot histograms of these distributions, and note how their shape differs from that of previous histograms. Calculate the mean of each distribution. Which species is the commoner? Assuming that the three occasions on which seven daisy plants were ringed represents a significant minor peak on the histogram, how could you explain its occurrence? Given that the area of the ring was 50 cm², and the area of the lawn was 200 m², calculate (to the nearest hundred) the number of plants of each species present on the lawn.

(Answers on p. 206)

Measuring the spread of a distribution

Having determined the mean, or in one of the other ways fixed the position of the peak of the distribution, the next step is to measure how widely the observations are scattered. This is called the **spread** or **dispersion** of the distribution; roughly, it corresponds to the width of the histogram.

There are several ways of measuring spread:

Range: This is the difference between the highest and the lowest values of the variable. In the example (Table 3, p. 9), the range is

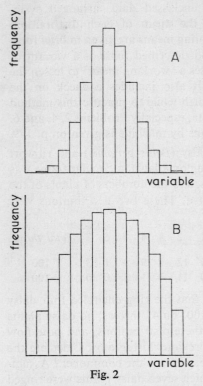

Fig. 2

from 28 to 165, so the range is 137 flowers. The range is very easy to calculate, but its usefulness is limited. Two quite different distributions may have the same mean and range (Fig. 2), but in one of them (A) the observations are much more clustered around the mean than in the other (B). Another disadvantage of range is that it is greatly affected by the extreme values of the variable. Suppose that the exceptional plant with 165 flowers had not been grown. The range would then have been only 116 flowers. The value of the range is too much affected by the few atypical observations which occur at the extremes of the distribution.

Semi-interquartile range: This eliminates the disadvantage of the simple range, by placing less emphasis on the extremes. A line through the median of a histogram divides it into halves of equal area; similarly, lines through the quartile points divide these halves

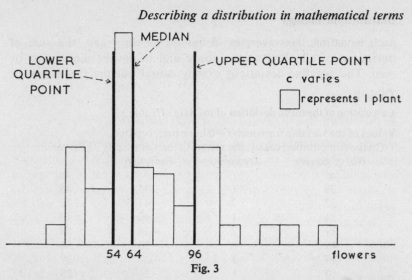

Fig. 3

into halves again, thus quartering the area of the histogram. This is
shown in Fig. 3, where the distribution has been quartered like this:

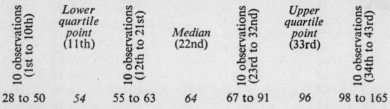

10 observations (1st to 10th)	Lower quartile point (11th)	10 observations (12th to 21st)	Median (22nd)	10 observations (23rd to 32nd)	Upper quartile point (33rd)	10 observations (34th to 43rd)
28 to 50	*54*	55 to 63	*64*	67 to 91	*96*	98 to 165

The interquartile range is measured from the lower quartile point
to the upper quartile point, and the *semi*-interquartile range is half
of the interquartile range. In the example, the interquartile range is
from 54 to 96, a difference of 42 flowers. The semi-interquartile range
is 21 flowers. This gives a better estimate of spread than the more
simply calculated range, since extremes do not play so large a part
in determining its value.

Mean deviation: Every value of the variable differs from the sample
mean by an amount which is called its **deviation**. The deviation (d) of
an observation (x) is given by the equation:

$$d = x - \bar{x}$$

If the observation is greater than the mean, the deviation is
positive. If it is less than the mean, the deviation is negative. One
could try to use the deviation as a measure of dispersion, by calculating the mean of the deviations of all the observations. This would
involve calculating Σfd, in which expression f is the frequency of

21

Statistics for biology

each deviation. However, by definition of the mean, the sum of deviations of all observations above and below the mean *must* be zero. The negative deviations exactly cancel out the positive ones.

Table 8

Calculation of the mean deviation of the data of Table 3

Values of the variable for which $f = 0$ have been omitted.
To make computation easier, the mean is taken as exactly 73 flowers.

No. of flowers	Frequency	Deviation	
x	f	d	|fd|
28	1	−45	45
32	2	−41	82
33	1	−40	40
39	2	−34	68
45	2	−28	56
46	1	−27	27
50	1	−23	23
54	1	−19	19
55	1	−18	18
56	2	−17	34
58	1	−15	15
59	1	−14	14
60	1	−13	13
61	1	−12	12
62	1	−11	11
63	2	−10	20
64	1	− 9	9
67	1	− 6	6
70	1	− 3	3
72	1	− 1	1
73	1	0	0
76	2	+ 3	6
85	2	+12	24
87	1	+14	14
91	1	+18	18
96	1	+23	23
98	1	+25	25
99	1	+26	26
104	3	+31	93
109	1	+36	36
111	1	+38	38
131	1	+58	58
144	1	+71	71
165	1	+92	92
	$\Sigma f = 43$		$\Sigma\|fd\| = 1040$

If Σfd is always zero, it is no use for measuring dispersion. This difficulty may be overcome by using $\Sigma|fd|$. This is calculated by summing all the quantities, fd, as before, but now ignoring their sign; negative deviations are considered as if they were positive.

The calculation of mean deviation is shown in Table 8. It has been worked in full, as an illustration of the principle. In practice, it would be done by a short-cut method, similar to that on p. 158. Summing the 2nd and 4th columns gives the two required quantities, $\Sigma|fd|$ and Σf. From these the mean deviation is calculated:

$$\text{Mean deviation} = \frac{\Sigma|fd|}{\Sigma f} = \frac{1\,040}{43} = 24 \text{ flowers}$$

Variance: The variance, too, depends on deviations from the mean, but to calculate it the deviations are squared before summing them. Like the squares of positive quantities, the squares of negative quantities are positive. Thus, the quantities to be summed are all positive, and there are no complications arising from negative deviations. Squaring the deviations has other advantages, which are even more important, since this operation gives the variance special algebraic properties which allow it to be used for many statistical procedures.

The variance is the mean of the *squares* of deviations, as given by the equation:

$$\text{Variance, } s^2 = \frac{\Sigma fd^2}{\Sigma f}$$

Variance is denoted by the symbol, s^2. This is a reminder that variance is a mean of squares. An example of the calculation of s^2 is given in Table 9, the first three columns of which are identical with

Table 9

Calculation of the variance and standard deviation of the data of Table 3

Values of the variable for which $f = 0$ have been omitted.
To make computation easier, the mean is taken as exactly 73 flowers.

No. of flowers	Frequency	Deviation		
x	f	d	d^2	fd^2
28	1	−45	2025	2025
32	2	−41	1681	3362
33	1	−40	1600	1600
39	2	−34	1156	2312
45	2	−28	784	1568
46	1	−27	729	729

Table 9 (*continued*)

No. of flowers	Frequency	Deviation		
x	f	d	d²	fd²
50	1	− 23	529	529
54	1	− 19	361	361
55	1	− 18	324	324
56	2	− 17	289	578
58	1	− 15	225	225
59	1	− 14	196	196
60	1	− 13	169	169
61	1	− 12	144	144
62	1	− 11	121	121
63	2	− 10	100	200
64	1	− 9	81	81
67	1	− 6	36	36
70	1	− 3	9	9
72	1	− 1	1	1
73	1	0	0	0
76	2	+ 3	9	18
85	2	+ 12	144	288
87	1	+ 14	196	196
91	1	+ 18	324	324
96	1	+ 23	529	529
98	1	+ 25	625	625
99	1	+ 26	676	676
104	3	+ 31	961	2883
109	1	+ 36	1 296	1 296
111	1	+ 38	1 444	1 444
131	1	+ 58	3 364	3 364
144	1	+ 71	5 041	5 041
165	1	+ 92	8 464	8 464
	$\Sigma f = 43$			$\Sigma fd^2 = 39\,718$

those of Table 8. Summing the second and fifth columns gives the required quantities, Σf and Σfd^2. Then the variance may be calculated:

$$s^2 = \frac{39\,718}{43} = 924 \text{ square flowers}$$

The variance has been calculated by squaring numbers which represent flowers; the variance is therefore expressed in squared units. Square flowers are certainly a mathematical abstraction!

To be more accurate, it is the *numbers* of flowers which have been squared, not the flowers themselves, and the units of variance are 'numbers-of-flowers squared'.

Standard deviation: This is derived from the variance, and does not involve square flowers or other esoteric units. The standard deviation is the square root of the variance, and therefore has the same units as the original observations. This makes it possible to compare the standard deviation with the mean, both being expressed in the same units.

In mathematical terms:

$$\text{standard deviation, } s = \sqrt{\left(\frac{\Sigma f d^2}{\Sigma f}\right)}$$

In the example (Table 9),

$$s = \sqrt{\left(\frac{39718}{43}\right)} = \sqrt{924} = 30 \cdot 4 \text{ flowers.}$$

As might be expected, the standard deviation of this example is greater than the mean deviation (p. 23), because proportionately more weight has been given to those observations which deviate more from the mean. Standard deviation and variance are more complicated to calculate than mean deviation, but because of their special statistical properties they are more useful measures of dispersion. From now on, only the variance and standard deviation will be used.

Another way of calculating variance and standard deviation

For each observation in a set of data there is a deviation, d, where, as defined on p. 21:

$$d = x - \bar{x}$$

From this definition it follows that:

$$d^2 = (x - \bar{x})^2 = x^2 - 2x\bar{x} + \bar{x}^2$$

This equation is true for each observation, and both sides of the equation may be summed for all observations without affecting the equality, giving:

$$\Sigma d^2 = \Sigma x^2 - 2\Sigma x\bar{x} + \Sigma\bar{x}^2$$

Now the mean, $\bar{x}$, is a constant for all items of the summation, so it may be transferred outside the summation sign of the second term on the right of the equation. Also, $\Sigma\bar{x}^2$ is simply $\bar{x}^2$ added to itself n times (once for each observation), so instead of $\Sigma\bar{x}^2$, write $n\bar{x}^2$. The equation then becomes:

$$\Sigma d^2 = \Sigma^2 x - 2\bar{x} \cdot \Sigma x + n\bar{x}^2$$

B

By definition of the mean, $\bar{x} = \dfrac{\Sigma x}{n}$, and the equation may be written:

$$\Sigma d^2 = \Sigma x^2 - 2 \cdot \frac{\Sigma x}{n} \cdot \Sigma x + n \cdot \frac{(\Sigma x)^2}{n^2}$$

$$= \Sigma x^2 - \frac{2 \cdot (\Sigma x)^2}{n} + \frac{(\Sigma x)^2}{n}$$

$$\therefore \; \Sigma d^2 = \Sigma x^2 - \frac{(\Sigma x)^2}{n}$$

The term, Σd^2, is the sum of squares of deviations from the sample mean. It is the same term as $\Sigma f d^2$, used in the calculations on p. 24. There the frequency (f) was included in the term to indicate that, in the special method of computation used, each value of d had been multiplied by its frequency before summation. Of course the same procedure can be used in calculating Σx and Σx^2 when calculating Σd by the new equation, and one can if preferred rewrite this equation as:

$$\Sigma f d^2 = \Sigma f x^2 - \frac{(\Sigma f x)^2}{\Sigma f}$$

Whatever way it is written, it comes to the same thing; whether the observations, their squares and their deviations are summed separately, or whether each value of the variable (or its square, or its deviation) are multiplied by its frequency and the products summed, the same result is obtained in the end (p. 18). The equation given above, without the f's, is the more fundamental one, and easier to remember.

In the example, the *sum of squares of observations* is given by:

$$\Sigma x^2 = 28^2 + 32^2 + 32^2 + \ldots + 144^2 + 165^2 = 270034$$

The *sum of observations, squared,* is given by:

$$(\Sigma x)^2 = (28 + 32 + 32 + \ldots + 144 + 165)^2 = 3417^2 = 9903609$$

The number of observations, $n = 43$. These three quantities may now be used for calculating Σd^2:

$$\Sigma d^2 = 270034 - \frac{9903609}{43} = 270034 - 230316 = 39718$$

This is the same result as was obtained by the calculations of Table 9, but it has been arrived at without actually working out any of the deviations, thus saving both time and effort. The value of Σd^2 can now be used for calculating variance and standard deviation, as before.

Describing a distribution in mathematical terms

1–6 Find the range, lower and upper quartile points, semi-interquartile range, variance, and standard deviation of the distributions previously obtained from the data on pp. 13–15. Instructions for the calculation of variance and standard deviation will be found on pp. 157–60. Techniques for calculating machines are given on pp. 181–4. The answers are on p. 206.

4 Samples and Populations

In Chapter 3, two specially important quantities were calculated, which between them describe the distribution of the numbers of flowers on antirrhinum plants.

Mean: $\bar{x} = \dfrac{\Sigma x}{n}$ = 73·19 flowers (p. 17)

Standard deviation: $s = \sqrt{\left(\dfrac{\Sigma d^2}{n}\right)} = 30\cdot4$ flowers (p. 25)

There are also some histograms (Fig. 1), which show the general nature of the distribution. From these items of information, obtained from a sample containing 43 plants, how may one deduce the corresponding features of the population of red-flowered antirrhinum plants in general?

The total population will also have a mean and a standard deviation, which will describe the position and dispersion of its frequency distribution in the same way as the mean and standard deviation quoted above describe the frequency distribution of the sample. So that there will be no confusion, the mean *of the population* will be denoted by the Greek letter, μ (pronounced mu), and the standard deviation *of the population* by the Greek letter, σ (pronounced sigma). Likewise, the variance of the population will be denoted by σ^2 (sigma-squared). The number of individuals in the population may be represented by another Greek letter, ν (pronounced nu).

If it were possible to count the numbers of flowers on all the ν plants of the population (which it is not, since most of them have never been grown), μ and σ could be calculated *directly*, according to equations essentially the same as those above:

Mean: $\mu = \dfrac{\Sigma x}{\nu}$

Standard deviation: $\sigma = \sqrt{\left(\dfrac{\Sigma d^2}{\nu}\right)}$

Now consider what would be the appearance of a histogram

drawn to illustrate the distribution of the population. There would be several thousands or even millions of plants, so there would be no need to group the data. There would be hundreds of plants with 28 flowers, hundreds with 29 flowers, and so on, up to 165 flowers or more. All the gaps in the distribution would be neatly filled. The

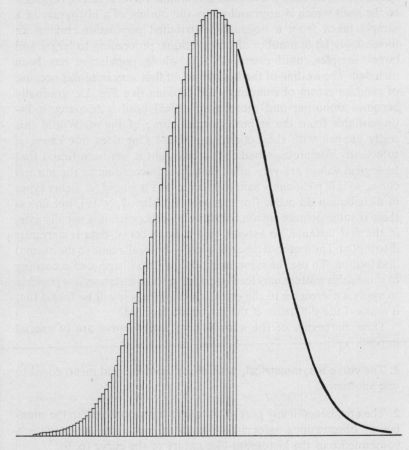

Fig. 4

chance variations which occurred with a small sample would be evened out, and the tops of the columns of the histogram would almost follow a smooth curve, rising symmetrically to a peak at the modal class (Fig. 4). The outline of the histogram would still consist of steps, but, with a very large number of observations, and the

29

smallest possible class interval (one flower), the outline of the histogram would approach very nearly to a smooth curve. The final shape of this curve would depend upon the characteristics of the antirrhinum population, but it is very likely that it would approximate to a curve of special shape, which has important statistical properties. This special curve is called the **normal distribution curve**. It can be regarded as the limit which is approached by the outline of a histogram of a sample taken from a normally distributed population, taking an increasingly large number of observations, proceeding to larger and larger samples, until eventually the whole population has been included. The outline of the histogram, at first very irregular because of random errors of sampling, and looking like Fig. 1A, gradually becomes smoother and more symmetrical until it becomes indistinguishable from the smooth normal curve of Fig. 4. Would this really happen with the antirrhinum data? One does not know, if sufficiently numerous, whether it would, but it has been found that biological values are very often distributed according to the normal curve, so it is reasonably safe to assume that it would fit. Other types of distribution do occur (for example, Problem 7, p. 19), but unless there is some definite reason for thinking otherwise, it is usually safe, in the first instance, to assume that a given set of data is normally distributed. The methods described in this book all relate to the normal distribution. To use the same methods for data distributed according to some other pattern *may* lead to error. In some instances it is possible to apply a correction to the error, and in others it will be found that it makes little difference if the errors are ignored.

Three properties of the normal distribution curve are of special importance:

1. The curve is symmetrical, with mode, median, and mean equal to one another.

2. The area beneath any part of the curve is proportional to the number of observations associated with that part. This property is a consequence of the histogram-like nature of the curve (p. 9).

3. Irrespective of (i) the scale on which it is drawn,
            (ii) the magnitude and units of its mean,
        and (iii) the magnitude and units of its standard deviation,

the curve has fixed mathematical properties which are susceptible to statistical analysis. Further, the independence of these properties from scale, magnitude, and units of measurement makes it possible to compile sets of statistical tables, like those at the end of this book, which do away with the need for drawing histograms for every set of data to be examined.

The derivation of a normal curve for a given mean and standard deviation is a very tedious calculation, but fortunately the tables make this unnecessary. The ways in which the tables can be used will be explained later, and for the moment discussion must return to the equations for μ and σ, given on p. 28. To calculate μ and σ, by means of these equations requires an observation from every member of the population, so that Σx, Σd^2, and ν may be known. Clearly, this is impossible, since only the 43 individuals of the sample are available, and it is on the data from these few that the estimates of μ and σ can be based.

The best estimate of μ is $\bar{x}$. One cannot do better than to take the mean of all available observations. This makes use of all the information provided by the sample, and there is no reason for supposing that μ has any other value which is different from $\bar{x}$. The value of $\bar{x}$ was derived from observations on 43 randomly selected and independent plants.

In estimating σ^2 and σ it is not possible simply to use s^2 and s. The reason for this prohibition is that the deviations of all the 43 observations are not completely independent. By definition, Σd^2 is calculated by summing all deviations; now suppose all deviations *except the last* have been summed:

$$-45 - 41 - 40 \ldots + 38 + 58 + 72 = -92$$

If the sum of deviations, so far, is -92, the next and final deviation *must* be $+92$ in order to fulfil the condition that the total deviation must be zero (p. 22). Thus 42 of the deviations are independent, and can take any value, but the 43rd is entirely dependent on the values of those which have come before it. This restriction on the 43rd deviation does not apply to any deviation in particular; in the example the deviations have been summed in the order of Table 9, but the summation can be in any order, and any one of the deviations might happen to be the 43rd *dependent* deviation. Whichever deviation is dependent does not really matter; the conclusion to be drawn from this discussion is that in a sample of size n, there are $(n-1)$ independent deviations upon which to base estimates of σ^2 and σ. To sum up, here are the equations for standard deviation derived so far:

B 2

$$s = \sqrt{\left(\frac{\Sigma d^2}{n}\right)}$$ The standard deviation of the sample.

$$\sigma = \sqrt{\left(\frac{\Sigma d^2}{\nu}\right)}$$ The *true* standard deviation of the population. The same equation as above, but with the symbols relevant to the population substituted. *Note:* The magnitude of ν will usually be far greater than that of n, and similarly that of Σd^2 will be far greater than that of Σd^2 in the equation above.

Estimated $\sigma = \sqrt{\left(\frac{\Sigma d^2}{n-1}\right)}$ Estimate of the standard deviation of the population Σd^2 and n have the same values as in the top equation.

To estimate σ, the number of truly independent deviations, $(n-1)$, is substituted in the usual equation for s. This requires further justification. In the note beside the middle equation it was pointed out that Σd^2 is usually very large; this is because it is the sum for all members of a large population. There is another difference between the Σd^2 of the middle equation and the Σd^2 of the other two equations. In the middle equation Σd^2 is calculated by summing squares of *deviations from the true mean* (μ) of the population, while in the other equations Σd^2 is the sum of squares of *deviations from the sample mean* ($\bar{x}$). In mathematical terms:

In the middle equation: $\Sigma d^2 = \Sigma(x - \mu)^2$

In the other two: $\Sigma d^2 = \Sigma(x - \bar{x})^2$

Now although $\bar{x}$ is the best available estimate of μ, it is unlikely that μ does in fact have exactly the same value as $\bar{x}$. Secondly, the observations of the sample are necessarily more closely grouped around *their own* mean ($\bar{x}$) than around any other value. In this discussion, it is important that they are more closely grouped around $\bar{x}$ than they are around μ. Deviations measured from μ would tend, on the whole, to be a little greater than deviations measured from $\bar{x}$. All the calculations are necessarily performed with deviations from $\bar{x}$ (which is known), whereas, to obtain an exact value for σ, they should be performed with deviations from μ. So, using s as an estimate of σ, with deviations which are slightly less than they should be, gives too low a value. To compensate for this, Σd^2 is divided by $(n-1)$, instead of by n. This gives a better estimate of σ^2 and σ because the compensation just allows for the under-estimation inherent in Σd^2.

Here, and on p. 31 there have been two differing explanations of the use of $(n-1)$ in estimating σ^2 and σ. Neither of them is entirely complete. They are two aspects of a principle, the core of which cannot be properly explained without recourse to mathematics. It is enough here to see that the operations of dividing by $(n-1)$ is reasonable, without worrying unduly about the underlying mathematics.

The principle referred to above is that known as **degrees of freedom**. In general, the number of degrees of freedom associated with a statistical quantity may be determined by taking the number of items (observations, deviations, means, etc.) from which the quantity is calculated, and subtracting the number of restrictions imposed upon the calculation. For example, the number of degrees of freedom associated with the mean, variance, and standard deviation of the sample of antirrhinum plants is calculated like this:

For the mean. There were 43 observations, randomly selected, without restriction. Therefore there are 43 degrees of freedom, and in estimating the population mean, Σx is divided by 43.

For the variance and standard deviation of the sample. There were 43 deviations randomly arrived at when picking out the 43 seeds for germination, so again there is no restriction, and there are 43 degrees of freedom. When calculating s and s^2, Σd^2 is divided by 43. The restriction that $\Sigma d = 0$ does not come into force until *after* the whole sample has been selected; only then, when selection is complete, do the values of $\bar{x}$ and of d become fixed. But this cannot happen until after the 43rd individual has been selected, so there can be no restriction on any individual value of d.

For the estimates of the variance and standard deviation of the population. There are 43 deviations, but by the stage one estimates σ^2 and σ, the sample has been chosen, and $\bar{x}$ already determined. This brings into force the restriction that $\Sigma d = 0$. Since there are 43 items and 1 restriction, this leaves only 42 degrees of freedom (43 − 1). In estimating σ^2 and σ, Σd^2 is therefore divided by 42.

Other examples of the calculation of degrees of freedom will be dealt with as they arise.

PROBLEMS

Estimate μ, σ^2, and σ for each of the distributions of Problems *1* to *6*, pp. 13–15. Techniques for computation are on pp. 154–60. The answers are on p. 207.

PROBABILITY

Probability can be expressed in a quantitative manner. If it is stated that the probability of an event is zero, it means that the event is impossible. Expressed, mathematically, the probability of the impossible event is given by $p = 0$. If an event is certain to occur, this is expressed by saying that $p = 1$. In this scheme, p always takes some value between 0, representing impossibility, and 1, representing certainty.

If a coin is tossed, it will come down either heads or tails and, supposing the coin is not biased, either event is equally likely to occur. Thus the probability of heads is 0·5, and the probability of tails is 0·5 also. The probability of the coin falling either heads or tails is $0·5 + 0·5 = 1$; it is *certain* to fall one way up or the other.

If two coins, A and B, are tossed together, there are four ways in which they may fall:

Way	1	2	3	4
Coin A	Head	Head	Tail	Tail
Coin B	Head	Tail	Head	Tail

Each of ways 1 to 4 is equally likely, and the probability of any one way is 0·25. The probability of all possible events is again 1 ($= 0·25 + 0·25 + 0·25 + 0·25$). The probability of getting one head and one tail, irrespective of which coin they appear on, is the probability of the occurrence of either way 2 or way 3, that is $0·25 + 0·25 = 0·5$.

In another instance, the probability of an event may be 1 in 20; if so, $p = 0·05$. The probability of this event *not* occurring is then 19 in 20, or $p = 0·95$, for the probability that the event will occur, plus the probability that it will not occur must always be unity, all possible events having been covered. In this example, $0·05 + 0·95$ gives the total probability unity; it is certain that the event either will or will not occur.

Another way of expressing probability in numerical form is as a percentage. Impossibility is expressed as 0%, and certainty as 100%. These figures are simply the values of p, multiplied by 100. When a single coin is tossed the probability of heads falling uppermost is 50%. If the probability of another event is given by $p = 0·05$, its probability may also be expressed as 5%, or 1 in 20. The probability of its not occurring is 95%.

This section is intended more to define the terms which will be used later, rather than to explain them at length. If the reader finds this section difficult to follow, it is recommended that a simple mathematical textbook should be consulted to provide further examples.

PROBLEMS

8 A dice, marked in the usual way, from one to six, is shaken repeatedly. What is the probability that a six will be shaken? Express the answer both in terms of p, and as a percentage.

9 With the same dice, what is the probability of shaking an odd number? an even number? Express the answers in terms of *p*, and as percentages.

10 A pack of 52 playing cards, containing the usual four suits, is shuffled and then cut. What is the probability of cutting an Ace?, the Ace of Diamonds?, any card of the club suit? Express all answers in terms of *p*, and as percentages.

Probability and the normal distribution curve

When the estimates of the mean (μ) and standard deviation (σ) of the population of red antirrhinums have been calculated it is possible to draw the normal curve to represent this population. The equation of the normal curve is a complex one, and plotting such a curve is tedious. It is shown in Fig. 5. The position of the peak of curve is fixed by the value of the mean (and also the mode and median), at 73·19 flowers (p. 17). The spread of the curve depends on the standard deviation. As explained on p. 31 this is estimated from the equation:

$$\sigma = \sqrt{\left(\frac{\Sigma d^2}{n-1}\right)} = \sqrt{\left(\frac{39718}{42}\right)} = \sqrt{(945\cdot7)} = 30\cdot75 \text{ flowers}$$

These estimates of μ and σ were substituted in the equation for the curve when drawing Fig. 5. The greater part of the area of the curve lies between CD and EF, that is, between 12·92 and 133·46 flowers. In fact, in the sample collected, 41 out of the 43 plants (=95·3% of the sample) were within this range. Considering the curve as a histogram, the area below any part of the curve is proportional to the frequency of observations having values within that part of the curve. For instance, the line AB divides the curve and the area beneath it into two equal parts. Half the observations are greater and half

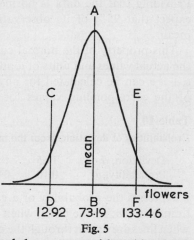

Fig. 5

are lesser than this median value, and the areas on either side of AB are equal. In terms of probability, for any given observation to fall on one particular side of AB, $p = 0\cdot5$.

It is convenient to take the whole area beneath the curve, including the area beneath the tails of the curve, which continue to infinity on either side, to be unity. Then the area under each half of the curve is 0·5, which has the same numerical value as p, the probability that any given observation will fall within that half of the curve. This property is true for other sectors of the curve as well. The lines CD and EF of Fig. 5 enclose 95% of the area beneath the curve. The remaining 5% of area is in the long tails to the left and right. If the total area of the curve is 1, the part between CD and EF has an area 0·95, and the probability of any given observation falling within this part of the curve has the same numerical value, so $p = 0·95$. In the example, 95·3% of the observations were within this part of the curve, which is good agreement with theory.

The values 12·92 and 133·46, which were used for positioning CD and EF, were calculated from two rather similar equations:

$$\mu - 1·96 \, \sigma = 73·19 - 1·96 \times 30·75 = \ \ 12·92$$
$$\text{and} \quad \mu + 1·96 \, \sigma = 73·19 + 1·96 \times 30·75 = 133·46$$

CD and EF have been located on either side of the mean at a distance of 1·96 times the standard deviation of the population. Now it is a property of the normal distribution curve that, no matter what the actual values of μ and σ may be, lines drawn at the positions $\mu - 1·96 \, \sigma$ and $\mu + 1·96 \, \sigma$ will enclose 95% of the area of the curve. Providing that the data is normally distributed, one can therefore expect that 95% of its observations will fall within 1·96 σ of its mean.

This property of the normal curve is independent of scale and of the actual values and units of μ and σ, and a relationship of the form $\mu \pm d \, \sigma$ can be calculated for any value of p, and tabulated. Some of the corresponding values for p and d are given in Table 10.

Table 10

Probabilities of deviations from the mean

Deviation, d	1·65	1·96	2·33	2·58	3·39
Probability, p	0·10	0·05	0·02	0·01	0·001

p expresses the probability of a given observation being *outside* the limits $\mu \pm d \, \sigma$. In the table, when $d = 1·96$, $p = 0·05$. This means that when lines are drawn through the curve at $\mu - 1·96 \, \sigma$ and at $\mu + 1·96 \, \sigma$, the probability of a given observation being outside these limits is 0·05, or 5%. In other words the probability of its being between the lines is 95%. This is the reasoning which led to the statements above, about the lines CD and EF. Similarly, the probability of an observation being less than $\mu - 2 \cdot 58\,\sigma$, or greater than $\mu + 2 \cdot 58\,\sigma$ is 0·01, or 1%. The farther the lines are placed from the mean, the less is the chance that a given observation will lie outside them. In terms of area, the farther apart the lines are drawn, the smaller is the area which lies outside them. As d increases, p decreases, which can be seen in Table 10.

As another example, the positions of the lines for the antirrhinum population can be calculated when $p = 0 \cdot 10$. The table gives $d = 1 \cdot 65$, when $p = 0 \cdot 10$, so the lines must be drawn at:

$$\mu - 1 \cdot 65\,\sigma = 73 \cdot 19 - 1 \cdot 65 \times 30 \cdot 75 = 22 \cdot 45$$
$$\text{and} \quad \mu + 1 \cdot 65\,\sigma = 73 \cdot 19 + 1 \cdot 65 \times 30 \cdot 75 = 123 \cdot 93$$

One would then expect to find that, *in the population*, 10% of plants would have fewer than 23 flowers, or more than 123 flowers; conversely, 90% would have between 23 and 123 flowers. As a check on this, note that in the actual sample 40 out of 43 plants ($= 93\%$) had between 23 and 123 flowers. This does not prove anything, since it was from the data of the sample that these estimates were derived, but one can at least see that the predictions have some resemblance to reality.

PROBLEMS

1–6 For $p = 0 \cdot 05$ and $p = 0 \cdot 01$, calculate the positions of lines to be drawn on the histograms of a few of the distributions from these problems. Answers are on p. 208.

The distribution of t

The calculations in the previous section gave a reasonably consistent result, but there are some flaws in the argument which were not then discussed:

1. The calculations were based on a small sample, whereas p is derived from the normal curve, which assumes an infinitely large population. With a small sample the theoretical probabilities will be less certainly realised.

2. Values of μ and σ are merely estimates, made from a small sample. The larger the sample, the more nearly will they approximate to the true values of μ and σ for the population.

Both these points underline the need to take into account the number of observations upon which an analysis is based. To take sample size into account requires the use of a quantity known as *t*. This is similar in many ways to the quantity *d*, found in Table 10; both are based on the normal distribution, but whereas *d* is simply the normal distribution of an infinitely large population, *t* takes into account the number of observations in the sample, and allows for the fact that the values of μ and σ are estimates, not their true, unknowable values.

The table of the distribution of *t* is given at the end of the book (Table 74, p. 192). The values of *t* have been calculated for the same values of *p* as in Table 10, but in Table 74 there are many rows corresponding with differing sample sizes. To be exact, it is not the sample size which influences *t*, but the number of degrees of freedom, and it has been shown (p. 33) that the two are closely related. Often, for a sample of size *n*, the number of degrees of freedom is $(n - 1)$, and this gives the point of entry into the table at the left-hand column.

In the bottom row of the table, for an infinite number of degrees of freedom (an infinitely large sample, equivalent to the whole population), the values of *t* are the same as those of *d* in Table 10. This is to be expected as *d* is based on the whole population. The farther up the table, the greater the values of *t*. The explanation of this is that, whereas with an infinitely large sample one can expect 95% ($p = 0.05$) of observations to lie within $1.96 \, \sigma$ on either side of the mean, with a sample of only 5 observations (and 4 degrees of freedom) μ and σ are less accurately estimated, and the lines must be drawn farther from the mean, $2.13 \, \sigma$ on either side, in order to contain 95% of observations.

For the antirrhinums the number of degrees of freedom is 42, and the nearest row of Table 74 is that for 40 degrees. In this row, when $p = 0.1$, $t = 1.68$, so:

$$\mu - 1.68 \, \sigma = 73.19 - 1.68 \times 30.75 = \quad 21.5$$
$$\text{and} \quad \mu + 1.68 \, \sigma = 73.19 + 1.68 \times 30.75 = 124.9$$

These figures indicate that 90% of the population will be found having between 22 and 124 flowers. Calculations based on *d* (p. 36) gave limits of 23 and 123, but it has now been shown that these limits were more precisely set than can be justified. Because the sample contained only 43 plants, and was not infinitely large, the limits at the same level of probability must be more widely set, at 22 and 124 flowers.

The distribution of *t* is of great practical value in statistical analysis. One example of its use has been given above, and several other applications will be given in the next section and later.

In general, t is expressed by:

$$t = \frac{\text{deviation from a mean}}{\text{standard deviation}}$$

In words, t is a deviation from the mean expressed in units of standard deviation. In the example above, t was 1·68, and the deviations of the limits were 1·68 units of standard deviation from the mean, at $\mu \pm t\,\sigma$.

PROBLEMS

1–6 Using the table of the distribution of t, calculate the more correct limit lines to be drawn through the histograms, for $p = 0·05$ and $p = 0·01$. Compare these results with those obtained from the original, but falsely based, calculations made after the previous section.

Many samples from a large population

One autumn, several thousand ripe capsules from radish plants were harvested in the course of a genetical experiment. Twenty-five samples, each consisting of 20 capsules, were chosen at random and the number of seeds in each capsule was counted. The frequency distribution for all 500 capsules is illustrated in Fig. 6 (continuous outline, unshaded). Let this be considered as a population, since it is based on a large number of capsules. The mean of this distribution, which will be referred to as the *grand mean*, is 6·04 seeds per capsule.

Next, the means of each of the 25 samples are calculated separately, and it is found that they range between 4·7 and 7·1 seeds per capsule. These 25 means are therefore more closely grouped around the grand mean than are the 500 individual observations, which range from 1 to 18 seeds per capsule. The effect is made clear in Fig. 6, where the distribution of the sample means is plotted in dashed lines, and stippled. In the diagram, the two distributions have been plotted on the same scale, and because they both represent a total of 500 capsules they both have the same area, but the distribution of individuals is broader and rises less to a peak than does the distribution of sample means. This result is not surprising, for in the process of taking means of samples, the extreme members of the population (having, say 1, 2, 10–18 seeds per capsule) lose their individual effect; the small tend to average out against the large. The chance of picking a sample with a majority of extremely large or extremely small numbers of seeds is remote. Most samples contain

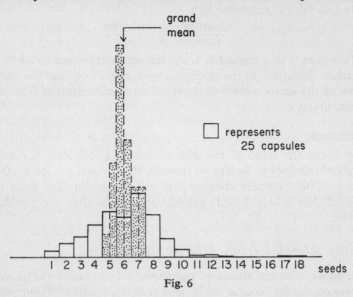

Fig. 6

an assortment of large and small capsules, and so sample means are fairly closely clustered around the grand mean of the population.

This effect can be expressed in statistical terms by saying that the standard deviation of the population (σ) is greater than the standard deviation of the sample means (σ_n).

The statement can be checked by calculation, by the usual method for the standard deviation of a population, giving:

$\sigma = 2.27$ seeds   (working with the 500 individual observations)

$\sigma_n = 0.67$ seeds   (working with the 25 sample means)

The standard deviation of the mean (σ_n) is a quantity which must be known before certain further calculations can be carried out. Here, having a large number of samples from a single population, it is possible to calculate it directly. However, in most experiments, such as the antirrhinum example, there is only *one* sample (the 43 plants) so it is not possible to calculate σ_n directly. Fortunately, it is possible to calculate σ_n indirectly by using an equation:

$$\sigma_n = \frac{\sigma}{\sqrt{n}},$$

in which *n* is the number of individuals in a sample.

For the radish data the equation gives:

$$\sigma_n = \frac{2 \cdot 27}{\sqrt{20}} = 0 \cdot 51 \text{ seeds}$$

This agrees reasonably well with the value 0·67 obtained by direct calculation, remembering that the equation is strictly true only for an infinitely large population and an infinitely large number of samples.

The equation shows that the magnitude of σ_n depends partly upon the number of individuals in the sample. The larger the number, the smaller is σ_n. The larger the sample, the less is the standard deviation of its mean, and so it lies closer to the grand population mean.

Calculating the limits of a mean

The mean number of flowers on the red-flowered antirrhinum plants has been shown to be 73 flowers, to the nearest whole number. This is a sample mean—how near to the population mean may it be expected to lie? The best estimate of the population mean is 73 (p. 31), but it would be remarkable if the actual value of μ turned out to be exactly equal to the mean of the sample. From this population it is possible to draw an infinite number of samples, with their means fairly closely clustered around the population mean, like the stippled histogram in Fig. 6. The antirrhinum sample is one of the infinite number of samples, and no further samples are available, so it is impossible to calculate the dispersion of this distribution of sample means directly, as was done for the radish data above. Instead there is the equation, which gives:

$$\sigma_n = \frac{\sigma}{\sqrt{n}} = \frac{30 \cdot 75}{\sqrt{43}} = 4 \cdot 689 \text{ flowers}$$

This is the estimate of the standard deviation *of the mean*. To connect this standard deviation with the probabilities of deviations from the population mean requires a value of t. At this point it is necessary to decide upon an acceptable level of probability. Absolute certainty (100% probability) cannot be obtained without access to the whole population, but for biological work 95% probability is usually satisfactory. Statements made as the result of the analysis will then be correct 19 times out of 20. Occasionally greater certainty is desired—perhaps before proceeding to treat humans with drugs which have just been tested experimentally—but too cautious an insistence on certainty can be a handicap to scientific progress. The degree of uncertainty (say 5%) has to be tolerated, but at least the statistician knows what level of probability may be attached to the results of an analysis, and will place in them neither more nor less confidence than they warrant.

Assume that the level of probability decided upon is $p = 0.05$ (5% chance of being wrong, 95% chance of being right). In the table of distribution of t (Table 74, p. 192), for 40 degrees of freedom (the nearest tabulated value to 42), and in the column headed $p = 0.05$, the value of t is 2.02. This implies that one will expect 95% of means of samples of 43 plants to lie within 2.02 units of standard deviation from the population mean. 2.02 units of standard deviation are $2.02 \times 4.689 = 9$ (to the nearest whole number).

This calculation is an application of the general equation for t given on p. 39:

$$t = \frac{\text{deviation from a mean}}{\text{standard deviation}}$$

or:

$$t \times \text{standard deviation} = \text{deviation from a mean}$$

In this application:

$$t \times \frac{\text{standard deviation}}{\text{of sample mean } (\sigma_n)} = \text{deviation from population mean}$$
$$2.02 \times 4.689 = 9$$

This figure indicates that whatever the actual value of the population mean may be, 95% of sample means will be within limits drawn nine flowers above it and nine below it. One cannot be *certain* that the mean of *this* sample is one of these 95%. There is no way of telling, but it is more likely that the sample is one of the majority. Not only can it be said that it is 'more likely', but a more precise statement can be made. It can be said that there is a 95% probability that the sample lies within nine flowers above or below the population mean, and only a 5% probability that it lies farther from the population mean than nine flowers. This is the probability level which can be attached to all following statements, resulting from the initial choice of probability level when consulting the table of t.

Now to locate the population mean. For the moment suppose that it has the value 68 flowers. Figure 7A shows the distribution curve for the *means* of samples. With $t\ \sigma = 9$ flowers, vertical lines are drawn at:

$$\mu - t\ \sigma = 68 - 9 = 59 \qquad \text{(lower limit)}$$
$$\mu + t\ \sigma = 68 + 9 = 77 \qquad \text{(upper limit)}$$

So, *if* the population mean is 68, 95% of sample means will lie between these limits, that is, between 59 and 77. Only 5% will be less

than 59, or more than 77 (the stippled area below the curve, and in
the tails beyond). The mean of the sample is 73. This is within the
limits, so the supposition that the population mean is 68 agrees with
the data obtained from the sample. It is being assumed, of course,

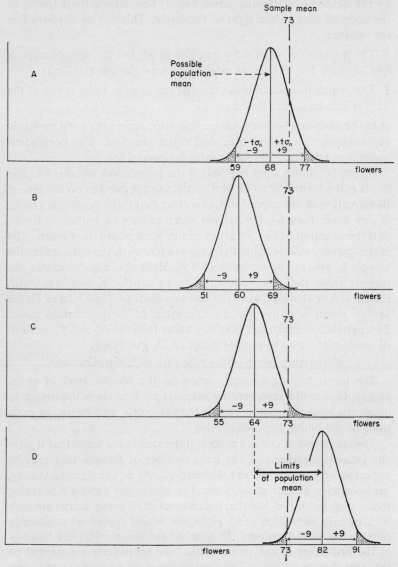

Fig. 7

that the sample is not one of the 5% of samples lying outside the limits.

Next, suppose that the population mean is 60 (Fig. 7B). As before, $t \sigma = 9$, so the limits are now 51 and 69. The sample mean (represented by the dashed line running down Fig. 7) lies outside these limits, in the stippled area to the right of the curve. This can be explained in two ways:

1. The population mean is 60, or close to 60, but the sample mean is one of those 5% of sample means which lie outside the limits.

2. The population mean is *not* 60, and the sample mean is one of the 95% which lie within 9 of it.

In the absence of other evidence one must accept the most probable explanation, the second one, and reject the first. The population mean is not 60; another value must be looked for.

Figure 7C shows what happens if the population mean is 64. The limits extend from 55 to 73, and so the sample *just* lies on the line of demarcation at the upper limit. It is clear that if the population mean is any lower than 64, the sample mean cannot lie within its limits, and the situation of Fig. 7B arises again. So, a population mean of 64 is the lowest possible value that is consistent with the data, *unless* the sample is one of the exceptional 5%. Similarly, Fig. 7D shows the highest value that the population mean can have, consistent with the data. Anywhere between the two positions of Figs. 7C and 7D the sample mean is within the required limits of the population mean. The population mean can take any value between 64 and 82, yet still be consistent with the sample mean of 73. Put briefly:

Population mean = 73 ±9 flowers (5% significance)

The term, 5% significance, refers to the chosen level of probability. One could have written instead: $p = 0.05$. It is important to quote the level of probability, for the value of t, and hence the positions of the limits, depends on this.

The statement above is a precise statement in the sense that it gives the greatest number and the least number of flowers that may be expected on average on red-flowered plants of the Eclipse variety, grown under similar conditions. The statement carries a warning that it may be untrue, but that the chances of its being untrue are only 5%, once in 20, which most biologists would regard as sufficiently reliable for most purposes. This statement summarises the findings of the first stage in the estimation of the reproductive potential of the antirrhinum plants. The next stage was to sample the ripe capsules and estimate the mean number of seeds found in each. The product of the mean number of capsules per plant, and the mean number of seeds per capsule would give the required estimate of the mean number of seeds per plant.

PROBLEMS

1–6 For each of the distributions, estimate the limits of the population mean, at the 5% significance level ($p = 0.05$). The table for t is on p. 192, a summary of equations is on p. 160, and the answers are on p. 208.

5 Comparing the means of two samples

In addition to the red-flowered antirrhinum plants in the garden, there were also 21 white-flowered antirrhinum plants, of the variety Sutton's *Intermediate White*. It was easy to see that they were different in several features other than the colour of their flowers. The white-flowered plants appeared to be bushier than the red ones, and though they had smaller flowers, they made up for this by having more flowers on each branch. It was decided to test whether this visual impression was substantiated by counting. The data were collected and analysed in the same way as was the data from the red-flowered plants. For convenience, the analysis for both varieties is summarised in Table 11. The figures in the table have been calculated

Table 11

Summary of data for red-flowered and white-flowered antirrhinum plants

Quantity	Symbol	Red-flowered	White-flowered
Number of plants in the sample	n	43	21
Mean number of flowers per plant	$\bar{x}$	72·76	169·50
Estimate of variance of population	σ^2	1046	3960
Estimate of standard deviation of population	σ	32·35	62·93
Range—lowest point		28	66
Range—highest point		165	266
Sum of squares of deviations from mean	Σd^2	43944	79200
Degrees of freedom	$n-1$	42	20

after grouping the individuals into classes, with a class interval of 20 flowers. This was done to save labour in computation, since a calculating machine was not available. The technique is described on pp. 155–7. Grouping has given values of $\bar{x}$, σ, σ^2, and Σd^2 which differ slightly from those used in previous calculations on the red-flowered plants, but these differences are relatively unimportant.

The estimated normal curves of these two distributions are shown in Fig. 8, though this has been done solely for illustrating this discussion, and does not form a part of the normal procedure of analysis.

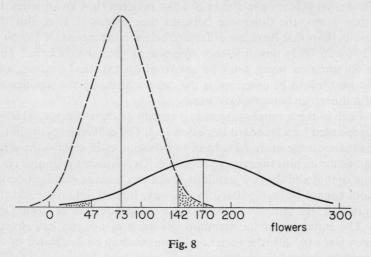

Fig. 8

There is a considerable difference between the means of these samples, though their ranges overlap. It is known that these samples are different in that one consists of red-flowered plants, and the other of white-flowered plants. Do they also differ in the mean number of flowers found on each plant? With respect to flower-number, are they from different populations, with different means, or are they merely two different samples selected from a single population? Two samples from a single population would probably have different means—how different do they have to be before it becomes clear that they are *not* from a single population? In everyday terms, do red-flowered plants have more flowers, on average, than white-flowered plants?

The original histograms, from which these spuriously smooth curves were drawn, were very irregular (Fig. 1D), especially that for the white-flowered plants. It is not possible to answer any of the questions above by merely examining the histograms. A more rigorous analysis is needed. The first step of this is to set up what is called a **null hypothesis**. This hypothesis is set up especially for the analysis, and can be tested statistically. It need have no direct connexion with any of the experimental hypotheses. After predicting what results would follow if the null hypothesis is true, the actual results are compared with the predictions. The null hypothesis is then either accepted or rejected, or perhaps accepted with caution.

In the example, a convenient and testable null hypothesis is that there is *no difference* in flower number between the two varieties. In other words, the difference between their means is zero. But the results show that there is a difference between the means, of 169·50 − 72·76 = 96·74. Is this difference *significantly* greater than zero? This is the question which must be answered by statistical analysis, and the next test to be described is the one for assessing the significance of a difference between two means.

Each of the means is subject to a certain amount of error, which is represented by a standard deviation (σ_n). The difference between the means must therefore be subject to two sources of error—the error of one mean, plus the error of the other. To calculate the double error one cannot add the two standard deviations because standard deviations cannot be manipulated in this way. Instead, one must add the variances; the sum will be the variance of the difference of means (σ_d^2).

The equation for the standard deviation of a mean has already been given (p. 40); the variance of a mean may be calculated by an equation derived from this:

$$\sigma_n^2 = \sigma^2/n$$

Using the suffix $_1$ to indicate the first sample (red-flowered), and the suffix $_2$ to represent the second sample (white-flowered), the addition of the variances of the two means may be written:

$$\sigma_d^2 = \frac{\sigma_1^2}{n_1} + \frac{\sigma_2^2}{n_2}$$

The square root of σ_d^2 is the required standard deviation of the difference of means (σ_d)

In the example, using data from Table 11:

$$\sigma_d^2 = \frac{1046}{43} + \frac{3960}{21} = 24\cdot33 + 188\cdot57 = 212\cdot9$$

$$\sigma_d = \sqrt{212\cdot9} = 14\cdot59 \text{ flowers}$$

Thus the difference between the means (= 96·74 flowers) has a standard deviation of 14·59 flowers. These two values can be substituted in the general equation for t (p. 39):

$$t = \frac{\text{deviation of the } \textit{difference} \text{ of the means, from zero}}{\text{standard deviation of the difference of means}} = \frac{96\cdot74}{14\cdot59} = 6\cdot63$$

The null hypothesis states that the difference of means is zero, but calculation shows it to be 96·74, in other words it deviates from the value expected on the null hypothesis by the amount 96·74. This amount of deviation is over six times its standard deviation. Is this greater than may be expected? The table of the distribution of *t* gives the maximum values of *t* that may be expected with a normally distributed population. In Table 74, for 60 degrees of freedom (nearest to 62), and $p = 0.05$, $t = 2.00$. One may expect that the difference between means will be up to but not exceeding *twice* its standard deviation in 95% of cases. In the example the difference is over *6 times* its standard deviation, which is far greater than would be expected. The null hypothesis requires that the difference of means shall be no greater than $2.00 \times 14.59 = 29.18$. The actual value is too great to be consistent with the null hypothesis, which must therefore be rejected. The difference of means is not *zero*, in other words, there *is* a significant difference between them. Using the null hypothesis, it has first been stated that the means were not different, and then shown that this statement is inconsistent with the data, proving that there is in fact a significant difference between the means.

The comparison of means may be treated in another fashion, like this:

The difference of means must lie between
$$96.74 - 2.00 \times 14.59 = 67.56$$
$$\text{and} \quad 96.74 + 2.00 \times 14.59 = 125.92$$

Following the same method as on p. 36.

The range of the difference of means, for $p = 0.05$, is from 67·56 to 125·92; this range does not include zero, so the null hypothesis is inconsistent with the data, and must be rejected.

Before continuing this example, it must be explained how the number of degrees of freedom was arrived at. Taking the two samples together, there were $43 + 21 = 64$ observations. There were two restrictions; for the mean of red-flowered plants, Σd^2 must equal zero, and the same restriction applied also to the white-flowered plants. So the number of degrees of freedom was $n_1 + n_2 - 2 = 64 - 2 = 62$.

In the table for *t* for 60 degrees of freedom (nearest to 62), and for $p = 0.001$, $t = 3.46$. Even this value is exceeded by the calculated value of *t* (6·63), so one may reject the null hypothesis with a 99·9% chance of being correct. This is a very high level of significance, amounting almost to certainty. If there were really no difference between mean flower numbers, and one repeatedly and randomly selected pairs of samples (one sample of every pair containing 43 plants, and the other sample containing 21 plants), one would expect to select more than a thousand pairs before finding a pair which showed as great a difference in their means as these two samples have shown. It is unlikely (though never forget that it is *possible*) that the pair of samples selected for this investigation is the one pair in a thousand which will lead to an unjustifiable rejection of the null hypothesis.

On the other hand, the one chance in a thousand can happen—it should do so once in every thousand occasions—so the null hypothesis *may* be correct, after all. Statistical analysis cannot tell if this is that rare occasion. Neither can analysis explain why the mean flower numbers differ, or determine what *biological* significance may be attached to this fact. Hypotheses about these aspects of the investigation are not within the scope of statistical analysis.

PROBLEMS

2, 4, 5, and *6* In each of these problems there are two distributions. For each problem, compare the means of the distributions, and assess the level of significance of the differences between them. Complete the analysis with a verbal statement of the conclusions which may reasonably be drawn.

11 In an investigation of the rate of germination of weed seedlings in various habitats, the soil temperature at a depth of 4 cm was measured on a sunny day in April. At five locations in a sunny garden bed, the temperatures were 14, 15, 12, $12\frac{1}{2}$, and $13°$ C. At five locations in the shade of evergreen trees, the temperatures were $9\frac{1}{2}$, 10, 10, 10, and $10°$ C. Calculate the mean temperature in the garden bed and in evergreen shade, and the 5% limits for each mean. At what level is the difference between these means statistically significant? Complete the analysis with a verbal statement of reasonable conclusions.

12 As an exercise in the computation technique given on pp. 155–7, use the data of Problem 2, grouped in the classes 5–9, 10–14, 15–19, 20–24, 25–29, and 30–34 mm, as for the histogram; calculate the mean of each distribution, its 5% limits, and assess the significance of the difference between the means.

13 Use the computation technique given on pp. 155–7 to work this problem. The tops were cut off 22 broad bean plants, four weeks old, at a level just above the two scale leaves near the base of the stem. The cut surfaces of the stems of 11 of the plants were smeared with a lanoline paste containing indole acetic acid (IAA), to see if the presence of this hormone would inhibit the development of the lateral buds in the axils of the scale leaves. The stumps of the other 11 plants were smeared with plain lanoline, as a control treatment. After two weeks the lengths of the axillary buds were measured; below are the data for the upper buds:

Bud length (mm)	Frequency IAA	Control
0– 39	3	3
40– 79	5	1
80–119	1	2
120–159	2	4
160–199	0	1
Total	11	11

For each treatment, calculate the mean bud length, and its 5% limits. Are the means significantly different?

14 In an investigation of abnormal human males, having a chromosome complement including XXY or XXXY instead of the usual XY, it was suggested that the abnormality tended to occur more in children born to older than average parents. Case-histories of 20 abnormal males were examined, and the ages of the parents at the time of birth of the child were discovered. The ages of the 20 mothers were:

31 21 29 34 41 43 27 39 38 37 30 39 16 28 32 21 31 21 45 28

The ages of 19 of the 20 fathers were:

33 22 30 45 43 36 27 35 34 52 38 55 20 34 26 25 29 48 32

Calculate the mean maternal age, and the mean paternal age, giving the 5% limits for the means. The mean age at which mothers in the general population bear children is 28·04 years; are the mothers of the abnormal males significantly older than mothers of the general population? Similarly, compare the mean age of fathers of abnormal males with the mean age of fathers in the general population, which is 31·04 years. In comparing sample means with general population means, which are gathered from an extremely large sample, one may assume that the latter mean is known exactly; therefore, $\sigma_d = \sigma_n$, where σ_n is the standard deviation of the sample mean.

This completes the part of the book dealing with basic statistical ideas. The remainder of this chapter, the next three chapters, and Chapter 10, deal with rather more specialised tests, and may be omitted on first reading. A different kind of test is dealt with in Chapter 11 and an outline of the nature and applications of tests is given in Chapter 12. It is recommended that the reader, who has decided to omit the next three chapters and Chapter 10 for the time being, should next read Chapters 11 and 12, followed by Chapter 9.

Differences between standard deviations

The reliability of the *t*-test as given above depends upon the two samples being drawn from populations which have identical or very similar standard deviations. Table 11 shows that this is not so for the two samples, and to perform the analysis under these circumstances may give misleading results. In a clear-cut example, like this one, when the calculated value of *t* is several times the tabulated value, there is no need to worry about this point. In the borderline cases, where the calculated *t* lies close to the tabulated *t*, one risks accepting a result as significant when it is not, or the other way about. A way of testing whether the standard deviations are near enough in value is given in the next section. If this test shows that the standard deviations are widely different from one another, the results of the *t*-test for the difference of means must be regarded with caution. More advanced statistical tests are available to deal with instances of this kind, and are described in some of the books listed on p. 205.

Limits for standard deviation and variance

The reference to differences between standard deviations leads to a new point of difference between the two varieties of antirrhinum. The distributions of Fig. 8 show a marked difference in spread. Do white-flowered plants really show a wider range of variability than red-flowered plants?

The variability is estimated by σ, and Table 11 shows that this quantity is about twice as big for white-flowered plants as it is for red. For a *large* sample, with n observations, the standard deviation (σ_σ) of its standard deviation (σ) is given by:

$$\sigma_\sigma = \frac{\sigma}{\sqrt{(2n)}}$$

For the red-flowered antirrhinums (see Table 11):

$$\sigma_\sigma = \frac{32 \cdot 35}{\sqrt{86}} = 3 \cdot 488$$

The table of t gives, for 40 degrees of freedom, and $p = 0.05$, $t = 2.02$. Therefore σ lies between

$$\sigma - t\sigma_\sigma = 32.35 - 2.02 \times 3.488 = 32.35 - 7.05 = 25.30$$
$$\text{and} \quad \sigma + t\sigma_\sigma = 32.35 + 2.02 \times 3.488 = 32.35 + 7.05 = 39.40$$

A similar calculation for the white-flowered antirrhinums shows that their standard deviation lies between 42.64 and 83.22. At the 5% level of significance these ranges do not overlap, so the standard deviations are significantly different. In numbers of flowers per plant, red-flowered antirrhinums are significantly less variable than white-flowered antirrhinums.

Though the technique above is useful for working out the limits of σ, it has the disadvantage that it is not reliable if there are fewer than 30 observations in a sample. The white-flowered sample had only 21 observations, so the limits worked above may be in error. Again, with such a clear-cut distinction between the two ranges, one will not be seriously mistaken if this point is ignored. For samples which contain fewer than 30 observations, and with larger samples when one merely wishes to compare two standard deviations to establish that they are different, without working out their limits, one can use the **variance ratio** test.

To calculate the variance ratio (F) the larger variance is divided by the smaller variance. For the two samples of antirrhinums (Table 11):

$$F = \frac{3960}{1046} = 3.786$$

This ratio has two sets of degrees of freedom, $n_1 = 21 - 1 = 20$ for the larger variance, and $n_2 = 43 - 1 = 42$ for the smaller. On pp. 193–6, are Tables 75–78, giving the distribution of the variance ratio, calculated for the usual levels of significance, though since there are two entries for the two sets of degrees of freedom, there is a separate table for each level of probability. The variance ratio has been calculated for a normal distribution and indicates, at the given level of probability, what is the maximum expected ratio between the variances of two random samples of given size, supposing that both samples are drawn from a single population. Thus, if two very large samples are drawn, so that n_1 and n_2 are large (indicated by infinity in the tables), one would expect that, if they come from a single population, their variances would be equal to that of the population, and so the variance ratio would be one. This value is found at the bottom right corner of all the tables. With fewer observations in either or both of the samples there is a greater chance of discrepancies in their variances, even though they come from the same population. For this reason the value of F increases towards the left and towards the top of the tables. The values of F are greatest for $p = 0.20$ and least for $p = 0.001$, for one is prepared to allow a greater discrepancy in estimates of variance (and hence variance ratio) when working at a low level of significance than at a high one.

In Table 76 ($p = 0.05$), for $n_1 = 24$, and $n_2 = 40$ (the nearest tabulated values), $F = 1.8$. The interpretation of this is that with a single, normally distributed population, by taking lots of pairs of samples, one of every pair having 43 observations and the other having 21 observations, by calculating their variances, and hence their variance ratio, one would expect to get a variance ratio larger than 1.8 in only 5% of the pairs. The variance ratio of the two samples is 3.786, which is greater than 1.8, so this ratio is too high to be the result of random selection from a single population. It could only arise by selection of the samples from two different populations, with different variances and standard deviations. In Table 78 ($p = 0.001$) we find that $F = 3.0$. This too is exceeded by the calculated F, so the difference between the variances, and hence standard deviations of these two samples, is highly significant.

With this test there is an alteration of the probability levels quoted at the head of the tables. The tables of F are calculated on the assumption that one does not know which of the variances is the larger, that is, which corresponds to n_1 and which to n_2. By the method used for this test, one decides from the beginning that n_1 corresponds with the larger variance and n_2 with the smaller. The effect of this is to alter the values of p to double those given at the head of each table. The result was significant when F was taken from the table for $p = 0.001$. In giving the final result of this test, the level of probability is properly quoted as $p = 0.002$.

This highly significant result has an unfortunate consequence. The test for the significance of the difference between the means (pp. 48–49) depended for its validity on the equality of the variances of the two samples. It has now been shown that these variances are unequal, and so the test for the difference of means is not strictly valid. In this and other examples where this occurs, one must then accept the results of the simple t-test with caution. Fortunately, in the example the calculated t was far greater than the tabulated one, so that even a most cautious acceptance would still concede significance. More refined statistical tests can be used when variances are not equal, and these are described in the advanced texts. Actual calculation, using one of these techniques (Cochran's test, see reference, p. 205) showed that, for the difference of means to be significant for $p = 0.001$, the calculated value of t must exceed 3·81 (instead of 3·46, as obtained by the method of p. 49); the calculated value of t was 6·63, and is still highly significant. It is plain that even a highly significant difference of variance need not upset the significance of the difference of means, when this is highly significant.

The variance ratio technique is applicable to two samples; when there are three or more samples the variances and standard deviations may be compared by Bartlett's test, which is described by Snedecor (1956).

PROBLEMS

4 Calculate the standard deviations (σ_σ) of the standard deviations of the distributions of numbers of stamens and carpels. By calculating limits for σ for each distribution show that these standard deviations are significantly different at the 1 % level.

4, 5, 6, 11, 13, and *14* Apply the variance ratio test to these problems.

6 Comparing three or more samples

Previous chapters have described the techniques used for comparing two samples. The two samples can be two sets of observations, like the counts of antirrhinum flowers, or they can be the numerical results of two series of experimental treatments. When there are three or more sets of experimental or observational data to be compared, different techniques of analysis must be employed.

Table 12

Water lost from Cherry Laurel leaves (mg/cm^2) in three days

Leaf replicate No.	Surface covered with jelly			
	Neither	Top	Bottom	Both
1	86	41	25	13
2	108	44	35	11
3	118	40	37	13
4	79	52	26	13

Consider the results (Table 12) of an experiment designed to measure the amounts of water leaving the upper and lower surfaces of a leaf. Sixteen leaves were detached from a Cherry Laurel plant, and their areas were measured. They were then divided into four batches of four leaves, and each batch was given one of four treatments. One batch was left untreated ('neither'), the other three batches were smeared with petroleum jelly on either their top, bottom, or both surfaces. This experiment could have been performed with just four leaves—one for each treatment—but by performing each treatment four times it is possible to gauge the extent to which apparently similar leaves, identically treated, vary among themselves. This experiment has thus been **replicated** four times. Identical (or as nearly as possible identical) individuals which are given identical treatment are called replicate individuals or, for short, **replicates**.

The leaves were weighed, hung in a shaded place with free air circulation for three days, and weighed again. The losses in weight are recorded in Table 12. Microscopic examination of surface sections of leaves from this plant showed stomata to be present in the epidermis of the lower surface only. The upper epidermis had no stomata, and had a very thick cuticle. It is probable, however, that though

56

such a thick cuticle should be virtually waterproof, it may have small cracks and scratches through which water vapour may pass. Inspection of the results shows some differences between the effects of the experimental treatments, though there is not a wide difference between the effect of covering the top surface, and the effect of covering the bottom surface. A comparison of these two treatments is a particularly interesting one, and the results must be examined statistically.

It is possible to perform the usual *t*-test on these results, taking the means of treatments, two at a time; one could compare 'neither' with 'top', 'neither' with 'bottom', 'top' with 'bottom', and so on. This would require six separate *t*-tests—a lengthy exercise. In a more extensive experiment, with seven treatments, there would be 21 *t*-tests to perform. At the 5 % level of significance it would be *expected* that one of the tests (actually 1 in 20) would give a value of *t* in excess of that in the table, even if the two sets of results were *not* significantly different. To get over this difficulty, to save long calculations, and to secure certain positive advantages, a different test is employed, called the **analysis of variance**.

The mean of all the values in Table 12 is 46·3 mg/cm², and every value deviates from the mean by a certain amount, called its deviation. In the terms used on p. 21, it could be said that for each observation, x:

$$x = \bar{x} + d$$

Each observation may be considered as the sum of the mean (46·3) and a deviation. The mean sum of squares of all deviations is the variance of the set of results in Table 12. Now this variance may be analysed further, for the single term, d, may be split into two parts:

1. deviation due to the effect of the treatment; call this T

2. deviation due to random differences between leaves; for example, the four replicates of 'neither' treatment lost 86, 108, 118, and 79 mg/cm² respectively, presumably because no two leaves, and the treatment they receive, are exactly alike. This random deviation is the same sort of deviation as that referred to on p. 36 and elsewhere, and the term d will be retained for it.

The observations may now be considered as being formed from the sum of the mean and two independent deviations:

$$x = \bar{x} + T + d$$

Every observation in the 'neither' column of the table, having come from replicate leaves, will have a constant deviation due to treatment

C

(T_1), so observations in that column may be written:

$$x = \bar{x} + T_1 + d$$

Of course, d is different for each, giving rise to the variation *within* the treatment, mentioned above. Similarly, the observations in the 'top' column may be written:

$$x = \bar{x} + T_2 + d$$

The magnitude of the experimental effect is represented by the difference between T_1 and T_2. Inspection of Table 12 shows that T_1 must be larger than T_2, for all the observations in the 'neither' column are larger than those in the 'top', but it is not possible to say by how much they differ, for, superimposed on *between*-treatment differences, there are the *within*-treatment differences, caused by random variations in d.

The total variance of the whole set of results consists of variance due to d and to T. The purpose of the analysis is to sort out the variances separately, so that effects of treatment may be established; this technique is therefore called the analysis of variance. It might be asked why one does not analyse the standard deviation, instead. As mentioned on p. 48, variances can be added, but standard deviations cannot. This form of analysis depends upon the fact that the total variance of the whole set of data is the sum of the variance due to T, and the variance due to d.

As was possible for Σd^2, for a single set of observations (p. 25), it is also possible to derive a short and convenient equation for calculating sums of squares of T. The derivation of the equation is rather more complicated, but in the end one obtains something of the same general form:

$$\Sigma T^2 = \frac{\Sigma\ (\text{treatment totals})^2}{v} - \frac{(\Sigma x)^2}{n}$$

in which the treatment totals are the totals of the column of Table 12, and $v\ (=4)$ is the number of observations in each column. Note that the second term on the right is exactly the same as that in the usual equation for Σd^2.

The example of Table 12 is worked in full, with special techniques for checking the calculations, in Table 57, pp. 163–4. It is found that:

$$\Sigma x^2 = 51669 \qquad (=\text{A})$$

$$\frac{(\Sigma x)^2}{n} = 34318 \qquad (=\text{D})$$

$$\frac{\Sigma\ (\text{treatment totals})^2}{v} = 50460 \quad (=\text{B})$$

From these, the *total* sum of squares of deviations (deviations due to both T and d):

$$\Sigma(T+d)^2 = \Sigma x^2 - \frac{(\Sigma x)^2}{n} = 51\,669 - 34\,318 = 17\,351$$

Part of this is due to the effect of treatment (between-treatments sum of squares):

$$\Sigma T^2 = \frac{\Sigma \text{ (treatment totals)}^2}{v} - \frac{(\Sigma x)^2}{n} = 50\,460 - 34\,318 = 16\,142$$

So, of the total sum of squares (17351), most (16142) is due to treatment effects, while the remainder, or **residual** ($17\,351 - 16\,142 = 1\,209$) is due to random differences between leaves.

Each of these sums of squares has its own number of degrees of freedom, worked out in the usual manner:

Total SOS: n observations, 1 restriction ($\Sigma(T+d)=0$), giving $(n-1)$ degrees.

Between-treatments SOS: u treatments, 1 restriction (Σ (Treatment totals) $=\Sigma x$), giving $(u-1)$ degrees.

Within-treatments (residual) SOS: n ($=uv$) observations, u restrictions (that the total for each of the u treatments shall have a given value), giving $(uv-u)$ degrees of freedom. This is generally written as, $u(v-1)$.

The next step is to set out a table, analysing the variance into its two components, as in Table 58, p. 164. Note that the numbers of degrees of freedom add up to 15, $(n-1)$, showing that the partitioning of degrees of freedom is correct. The column on the right contains mean squares, obtained by dividing each sum of squares by its degrees of freedom. Thus the mean squares are similar to variances, in fact the residual mean square is the residual variance of the set of data. It gives a measure of variance due to error and random variation throughout the whole set of results. Since $\sigma^2 = 101$, this gives $\sigma = 10$. This value of σ can be used for calculating limits for the means of treatment totals, using the method outlined on pp. 41–5. There is only one provision, an assumption which underlies the whole analysis. It is assumed that the residual variance is the same for all treatments. If it is not, part of it will have been separated out along with the variance due to T, and the analysis will not be valid. Also, each treatment will have a different value of σ, and it will not be possible to calculate the limits of their means, using the same σ for all. For guidance on this point the advanced texts may be consulted, but unless the analysis is of critical importance it is not usually necessary.

It was said above that the mean squares are similar to variances, but the mean square for between-treatment effects is not simply a variance. It is the sum of the residual variance and v times the variance due to treatments. Knowing the value of the residual variance, and the value of v, it is possible to work out the value of the variance due to treatments. Occasionally this may be required, but usually it is sufficient to know whether or not this variance is significantly greater than the residual variance. If it is, the ratio:

$$\frac{\text{between-treatment mean square}}{\text{residual mean square}}$$

will be significantly greater than would arise by chance selection of observations from a normally distributed population. The statistical significance of this ratio may be tested by the variance ratio test (p. 53), using the tables of F (Tables 75–78, pp. 193–6).

In the example, the variance ratio, $F = 5381/101 = 53$. From Table 78, when $p = 0.001$, $n_1 = 3$, and $n_2 = 12$, it is found that $F = 10.8$. The calculated value of F is greater than the tabulated one, so the calculated value is significantly greater than may be expected in the absence of treatment effects. Therefore the between treatment mean square is significantly greater than the residual mean square, and it may be stated, with a probability of 0.1% of being incorrect, that the jelly treatment has a significant effect upon the rate of loss of water from the leaves. Note that in this application of the variance ratio test, the residual mean square is *always* the divisor, even if it is greater than the between-treatment mean square. In the same way, n_2 always is the number of degrees of freedom of the residual. This being so, the probability levels are as printed on the tables, not doubled as on p. 54. Of course, if the residual mean square was greater than the between-treatments mean square, one would not bother to calculate it and apply the test, for all values of F in the tables are unity or more, and it would be obvious without calculation that the treatment effect is not significant.

This analysis has shown that treatment, in general, has a significant effect on water loss. It has not shown that any given pair of treatments give significantly differing rates of water loss. It is probable that the overall effect of treatment is mainly the difference between 'neither' and 'both', and not greatly contributed to by the differences between 'top' and 'bottom'. This aspect of the analysis is dealt with further on pp. 103–10.

Now comes an interesting new development. At the beginning of the experiment, the areas of the leaves had been measured, so that water loss could be expressed in terms of unit area, and also to assist in sorting the leaves into four groups containing leaves of more or less equal size. Before allotting them to their treatment the leaves had been graded according to area. Grade 1 contained the four smallest leaves, grade 2 the next four larger leaves, grade 3 the next four larger, and grade 4 the four largest leaves. One leaf from each grade was allotted to each treatment. Thus the leaves were not identical replicates, as has been assumed so far, but differed in area. If the figures of Table 12 are rearranged according to the leaf-size grades, it gives Table 13.

Table 13

Water lost from Cherry Laurel leaves (mg/cm²) in three days

(data from Table 12, rearranged according to area of leaf)

Area of leaf	Surface covered with jelly			
	Neither	*Top*	*Bottom*	*Both*
1 (smallest)	86	40	26	13
2	108	44	25	13
3	118	52	37	13
4 (largest)	79	41	35	11

Inspection of these figures now suggests that, within each jelly treatment, the smallest and the largest leaves lose water at a lesser rate than those of medium area.

The deviation of each observation now appears to be made up of not two components, but three:

$$x = \bar{x} + T_j + T_a + d$$

What was previously called T is now T_j, to distinguish it as a treatment effect due to jelly treatment. What was previously d, is now split into T_a, an effect due to leaf area, and a random, residual error, which is still called d. In parallel with this, the total sum of squares of deviations may now be partitioned into three:

1. sum of squares due to differences between jelly treatments (T_j)

2. sum of squares due to differences in area (T_a)

3. the residual sum of squares due to random differences between leaves.

The analysis of variance will test whether either (1) or (2) or both, are significantly greater than (3). The working for this example is given in full on pp. 166-70. It follows the same pattern as the previous analysis, except that there are now totals of rows from which to calculate what portion of the total sum of squares is due to the effects of leaf area.

61

In this two-factor analysis there has been no increase in the amount of experimental work required. There are still only 16 leaves to be treated and weighed, yet there is a great gain in information. From one experiment it has been possible to gain information about the effect of jelly treatment *and* the effect of leaf area. Without the analysis of variance this would require two experiments. Further, as explained on p. 169, by separating each source of variation (treatment, area, and residual) the resolving power of the analysis has been increased, and greater reliance may be placed upon the final statements.

The analysis of variance is a powerful experimental tool; for further examples of its application, see pp.103–10, and consult the books listed on p. 205.

PROBLEMS

15 In an investigation of the effect of microhabitat on the germination of weed seeds, soil temperatures 4 cm below the surface were measured in five locations in each of three habitats:

Temperatures, in degrees Celsius

Sunny garden bed	Under evergreen shade	In shade of wall, soil covered by about 10 cm depth of dead lime leaves
14	10	9
15	10	9
12	10	10
12½	10	9
13	9½	8½

Perform an analysis of variance to show whether there are significant temperature differences between the three habitats.

16 In the same investigation, samples of soil were collected at the same locations. The samples were weighed, dried, and reweighed to determine their water content. The results were:

Water-content (%)

Sunny garden bed	Evergreen shade	Shade, under leaf litter
5	3	14
7	2	8
7	2	16
4	1	15
9	3	14

Perform an analysis of variance to show whether there are significant differences in the water content of the soil in the three habitats.

17 (a) Using the phenosafranine method,[1] the oxygen content of the water of a stagnant pool was measured at three depths, top (15 cm below the water surface), middle (half-way between top and bottom), and bottom (7·5 cm above the muddy bottom). These three measurements were repeated at two other sites in the pool. The readings obtained, in cm³ dissolved oxygen per litre of water, were:

Site	Top	Middle	Bottom
1	3·00	2·40	2·04
2	4·94	4·05	2·24
3	3·90	3·36	2·44

Ignoring differences between sites, that is, considering the three sites as three replicate readings of the three depths, perform an analysis of variance to show whether there are significant differences between the oxygen content at different depths.

(b) Carry out a two-factor analysis of variance on the above data, taking into account both depth and site. Comment on the levels of significance found for the effects of depth and of site.

[1] See Dowdeswell, W. H. (1959), *Practical Animal Ecology*, Methuen.

7 The correlation of two variables

Sometimes one may measure two connected variables, as in the antirrhinum example, where one could have counted the number of branches on each plant as well as the number of flowers. As well as possibly establishing that the white-flowered plants were more branched than the red-flowers plants, a simple comparison using only the single variable, branch-number, one could also have tried to find some relationship between these variables. Plants with the most flowers might have the most branches; it is also conceivable that they might have the least. In investigations involving measurements of physical quantities one may record two measurements for each individual. For example, a sample of broad beans (variety, *Roger's Emperor*) was examined, and for each bean, the length and the weight were measured and recorded (Table 14). To give a visual impression

Table 14
Weight and length of broad beans

Bean No.	Weight (g)	Length (cm)
1	0·7	1·7
2	1·2	2·2
3	0·9	2·0
4	1·4	2·3
5	1·2	2·4
6	1·1	2·2
7	1·0	2·0
8	0·9	1·9
9	1·0	2·1
10	0·8	1·6

of a distribution of this type (a *bi*variate distribution) one cannot use a histogram. Instead one uses a **scatter diagram** (Fig. 9). Each point on the diagram represents one bean. Its position on the diagram depends upon the magnitude of the two variables. The distribution of the points seems to show that the longest beans are the heaviest, and the shortest are the lightest—weight and length are **correlated**. In this example, correlation is positive—greatest weight is found with greatest length. In other instances, correlation can be negative (Fig. 10B). The birth-weight of pigs might be negatively correlated with the size of litter — the larger the litter, the smaller the birth-weight of the individual pigs of that litter. A worked example of negative correlation will be found on p. 119.

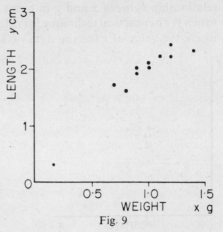

Fig. 9

A scatter diagram, like a histogram, is a convenient way of displaying results, but it does not give a reliable way of deciding whether there is a significant correlation. Figure 10c illustrates two variables with *no* correlation (*not* the same thing as negative correlation). Compare this with Fig. 10D and try to decide if there is correlation in the latter.

It is not easy to be sure about this; one must replace subjective examination of scatter diagrams with something more precise and impartial.

To measure correlation, statisticians use the **correlation coefficient,** *r*, which is given by this equation:

$$r = \frac{\Sigma((x - \bar{x})(y - \bar{y}))}{\sqrt{(\Sigma(x - \bar{x})^2 \Sigma(y - \bar{y})^2)}}$$

In this expression, one variable is represented by x, the other by y; $\bar{x}$ and $\bar{y}$ are the means of the two sets of variables. The terms $(x - \bar{x})$, and $(y - \bar{y})$ are deviations from the means as used in computing variance and standard deviation. The term $\Sigma((x - \bar{x})(y - \bar{y}))$ shows that, for each individual, one measures the deviation of x from the mean, $\bar{x}$, and the deviation of y from the mean $\bar{y}$, and multiplies these two deviations together. It will shorten the notation to use the symbol d to denote deviation, as in earlier chapters. Now to differentiate between deviations in x and deviations in y:

$$d_x \text{ stands for } (x - \bar{x}) \quad and \quad d_y \text{ stands for } (y - \bar{y})$$

Rewriting the equation for r in this notation gives:

$$r = \frac{\Sigma d_x d_y}{\sqrt{(\Sigma d_x^2 . \Sigma d_y^2)}}$$

The expression *above* the line gives a measure of the way x and y vary together. The expression *below* the line, containing the sums of

65

squares of deviations in x and in y, is related to the variances of x and y taken separately. As a whole, the equation for r expresses the relationship *between* x and y in terms of the variances *within* x and *within* y. The practical technique for calculating r is given on pp. 175–6, where the value of r for the data of Table 14 is obtained.

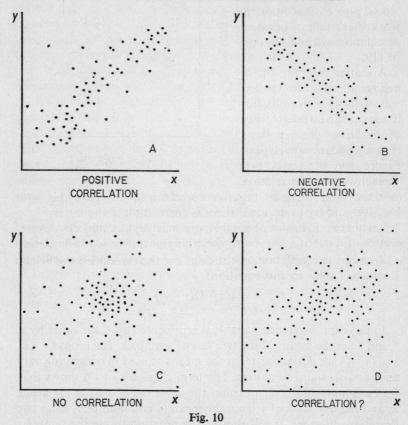

Fig. 10

The correlation coefficient can range between -1 and $+1$. If it is -1, correlation is perfect, and negative; a value of $+1$ means perfect positive correlation; zero means no correlation. Intermediate values of r indicate varying degrees of correlation, from very slight, when r is near to zero, to very strong, when r is near to -1 or $+1$. The significance of a calculated value of r may be gauged by comparing it with values from a table of r (Table 79). In the example (see calculations on p. 175), $r = 0.898$. The tabulated value of r for 9 degrees of freedom $(n-1)$, and $p = 0.1$, is 0.735. The calculated value exceeds this so one may say that there is positive correlation between the weight and length of these beans, with only a 1% chance of being wrong.

If correlation turns out to be significant, one may wish to derive an equation relating one variable to the other; this will give a line which can be drawn through the cloud of points on the scatter diagram. When correlation is perfect all points will lie on a single straight line, which can be drawn without the need for calculation. With the usual cloud of points it is not possible to manage without calculation, and there are *two* lines which may be drawn. These lines are called **regression lines**, and first to be considered is *the regression of y on x*, which is drawn in Fig. 11A. Here is the data from Table 14, plotted on a larger scale than in Fig. 9. Each of the points deviates from the regression line. In the figure, the deviations in the vertical direction (deviations in *y*) are indicated by the dashed lines. These lines represent the deviations in *y*, and the regression line has been drawn in such a position that the sum of the squares of the lengths of these lines is a minimum. To move the line up or down, or to alter its slope, will increase the sum of squares of deviations from the line. In this sense, this is the best line that can be drawn through the cloud of points, but there is another 'best line' shown as the thin, dashed line in Fig. 11B. This is the *regression of x on y*. This time, the deviations are measured horizontally (deviations in *x*), as represented by the thick, dashed lines. The position of the regression of *x* on *y* is such that the sum of squares of the deviations in *x* is a minimum.

The two regression lines do not coincide, unless correlation is perfect ($r = \pm 1$) for it is only then that deviations in *x* and *y* can be minimised simultaneously. When correlation is less than perfect, the lines cross, as in Fig. 11B, at a point called the **mean centre** of the distribution. The coordinates of this point are $(\bar{x}, \bar{y})$.

To plot regression lines one calculates $\bar{x}$ and $\bar{y}$. This gives the position of the mean centre, through which both lines pass. To obtain the slope, or gradient of the lines one calculates *b* for each line. For the regression of *y* on *x*, *b* is given by:

$$b = \frac{\Sigma d_x d_y}{\Sigma d_x^2}$$

Then substitute the calculated values of $\bar{x}$, $\bar{y}$; and *b* in the equation:

$$y = \bar{y} + b(x - \bar{x})$$

This gives the regression of *y* on *x*. It has been done for this example on p. 176, the final form of the regression equation being:

$$y = 1 \cdot 09x + 0 \cdot 93$$

67

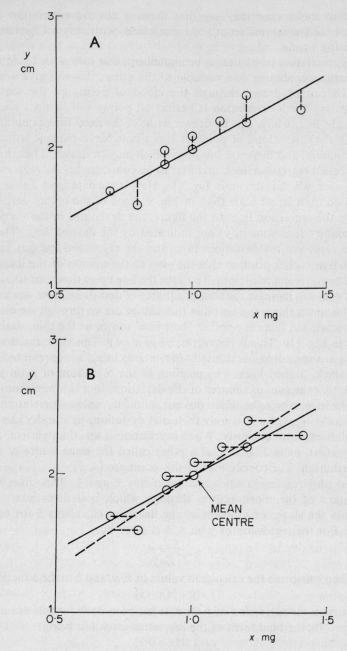

Figs. 11 A and B

This is the equation used for drawing the regression line on Fig. 11A. The regression of y on x can be used for estimating values of y from given values of x. Suppose one has a bean which weighs 1·3 g. Its most likely length is estimated by substituting $x = 1·3$ in the regression equation:

$$y = 1·09 \times 1·3 + 0·93 = 1·42 + 0·93 = 2·3 \text{ cm}$$

The regression of x on y is calculated in a similar way, using $b = \Sigma d_x d_y / \Sigma d_y^2$, and substituting this in the equation $x = \bar{x} + b(y - \bar{y})$. This has been done for the example on p. 176, giving the regression equation:

$$x = 0·74y - 0·49$$

This equation can be used for estimating values of x from given values of y. Suppose one has a bean which is 1·8 cm long. Its most likely weight is estimated by calculating:

$$x = 0·74 \times 1·8 - 0·49 = 1·33 - 0·49 = 0·84 \text{ g}$$

In these calculations from regression lines there are no fixed levels of significance. There are ways of determining statistical limits for b, and thus for deriving limits for values of x and y obtained from the regression equations, but these techniques are outside the scope of this book.

Throughout the analysis of this example it has been assumed that the two variables are linearly related, that the regression line is a straight one. Other sorts of relationship may be found in biological work. For instance, the number of bacteria in a culture increases in a geometrical progression as time passes, in the early stages at least. To deal with such correlations one uses extensions of the methods of this chapter. An example of this will be found on p. 123.

The existence of statistically significant correlation between two variables does not prove that variation in one *causes* variation in the other. A bean is not heavier *because* it is longer, or the other way about. Presumably both of these variables are governed by a third factor, which may be genetical, may be the result of the nutritional status of the plants from which the seeds were obtained, or may merely be due to the position of the beans within the pod, seeds at the ends of the pod being lighter and shorter than those in the middle. Any one or more of these factors may be affecting both length and weight, bringing about the correlation we have observed. Statistics cannot tell which factors affect which variables—only sound biological reasoning, with statistically significant backing, can explain these things.

PROBLEMS

4 The data for this problem are set out on p. 14, showing the number of stamens and the number of carpels for each plant. For example, the first plant has 60 stamens and 30 carpels; the second has 54 stamens and 21 carpels, and so on. Plot a scatter diagram to illustrate the relationship between the numbers of stamens and the numbers of carpels. Calculate *r*, and find the significance of the correlation. Use the data from all 51 flowers, or, to save arithmetic, use the data from the first 20 flowers only.

18 An investigator wished to know if the number of spines on the margin of a holly leaf was correlated with the size of the leaf. The lengths of 50 leaves were measured and the number of spines counted, omitting the terminal spine:

Length (mm)	73	45	60	68	63	65	75	60	73	75	53	53	33	59	54	55	54	53
No. of spines	9	11	15	18	18	20	14	10	17	12	5	11	2	13	13	12	10	14
Length (mm)	59	73	62	60	78	66	68	72	46	48	36	50	56	51	55	63	66	64
No. of spines	13	12	14	17	8	16	12	8	2	15	1	15	10	10	7	6	8	14
Length (mm)	56	61	61	52	68	39	48	56	65	39	31	50	56	53				
No. of spines	0	1	7	0	3	0	0	5	12	6	1	7	3	1				

Plot a scatter diagram to illustrate these data. Calculate *r*, and find the significance of the correlation. Calculate the equations for the regression of length on spine number, and of spine number on length. What is the most likely number of spines on a leaf 70 mm long? What is the most likely length of a leaf with six spines?

19 The amount of ascorbic acid present in a given volume of solution may be estimated by measuring the extent to which it decolorises the blue starch-iodine complex. The amount of decolorisation is measured with a photo-electric absorptiometer. To standardise the procedure a number of solutions were made up containing known amounts of ascorbic acid. The reading of the meter on the absorptiometer was taken as each of these solutions was placed in the apparatus:

| Amount of ascorbic acid (μg/cm^3) | 150 | 300 | 450 | 600 | 750 | 900 |
| Meter reading | 5·9 | 4·8 | 3·7 | 2·4 | 0·9 | 0·0 |

Calculate *r*, and find the significance of the correlation.
Calculate the regression of the amount of ascorbic acid on the meter
reading, to obtain the best calibration curve. A sample of milk was
tested for ascorbic acid; allowing for the turbidity of the milk itself,
the meter reading was 1·3. What is the best estimate of the ascorbic
acid content of the milk?

8 The chi-squared test

In a genetical experiment, some pure-breeding, red-flowered antirrhinum plants were crossed with some plants of a pure-breeding, white-flowered variety. The next generation (the F_1) all had pink flowers, since they were heterozygous for flower colour; they were allowed to interpollinate randomly. The F_2 generation then consisted of 145 plants with red flowers, 289 with pink flowers, and 138 with white flowers. It is interesting to test whether this result is in agreement with the ratio, *1 red : 2 pink : 1 white*, predicted by genetical theory.

Before performing the analysis, it is important to consider the nature of the data, which differs from that of previous examples. In all the examples analysed so far, the observations have been the values of one, or more, variables. In the genetical experiment just outlined, the variable is the flower colour—it does not have numerical values, either continuous, or discontinuous. Instead, it has three categories—red, pink, and white—and when recording the results of the cross, one counts the number of plants which fall in each category. There are 145 observations of the red category, 289 of the pink, and 138 of the white, making 582 observations altogether. Data of this sort, which consist of the numbers of individuals in given categories, are called **enumeration data**. The results of breeding experiments are frequently of this kind, so this chapter is especially important to those who work in genetics.

The fertilisation of the ovules of the pink-flowered F_1 plants, by the gametes from the pollen grains of other pink-flowered F_1 plants, is the outcome of a series of events, many of which are random. For any given ovule, the probability that it will be fertilised by a male gamete, bearing the gene for redness, is 0·5, and there is the equal probability that it will be fertilised by a gamete bearing a gene for whiteness, instead. Given large enough numbers, it is expected that the observed ratio of flower colours will be near to the *1 : 2 : 1* ratio predicted by theory, but since the events leading to fertilisation are effectively random, like the tossing of a coin, one does not in practice expect to obtain a ratio which is exactly *1 : 2 : 1*. How far may the observed ratio deviate from the predicted ratio before it may be safely assumed that the deviation is not solely due to randomness, but also to some flaw in the predictions?

If a coin is tossed 100 times, and it lands heads 45 times and tails 55 times, the probability of this occurrence can be calculated by an extension of the method used on p. 34, as described in mathematical textbooks. The probability of 45 heads and 55 tails is, $p = 0.049$, which may be compared with the probability of the most likely occurrence, 50 heads and 50 tails, for which $p = 0.080$, and with the probability of an unlikely occurrence, such as 20 heads and 80 tails, for which $p = 0.000000000426$. The probability of the last occurrence is so much less than that of the other two, that if 100 throws gave only 20 heads, one would strongly suspect that something was wrong with the coin, or the method of throwing it, these probabilities all having been calculated on the expectation that the probability of obtaining heads on a *single* throw is 0.5. The ratio, *20 heads : 80 tails*, deviates so far from the expected ratio, *50 heads : 50 tails*, that the hypothesis that heads and tails are equally likely must be rejected. What is now required is some measure for this deviation, so that it may be linked to statistical tests.

The measure of deviation of ratios is χ^2 (*chi*-squared, pronounced *ky*-squared) which is given by the equation:

$$\chi^2 = \Sigma \frac{(O - E)^2}{E}.$$

As an example of its calculation, take the figures given in Table 15, which refer to the breeding experiment mentioned above. The observed (O) and expected (E) numbers of plants are shown in the first two rows of the table. The expected numbers have been obtained by dividing the total number of plants according to the ratio predicted by theory. The prediction, in this example a *1 : 2 : 1* ratio, is the null hypothesis of the analysis. It has to be decided whether the observed ratio deviates significantly from that predicted by the null hypothesis. The magnitude of the difference is shown in the third row of the table ($O - E$). The total deviation is, of course, zero.

To find χ^2, each deviation is squared (4th row), divided by E (5th row), and then the quotients are summed. The quantity χ^2 has a distribution which has been computed and tabulated (Table 80). The table sets out, for several levels of probability, values of χ^2 which may be obtained when there are *only random deviations* between observation and expectation. Larger deviations give larger values of χ^2, and these are found towards the right side of the table, where probability levels are least. With random deviation one would expect greater deviations to occur least often. The magnitude of χ^2 is affected also by the number of terms to be summed, the number of categories. This is why χ^2 increases steadily down the columns of the table. Each row of the table relates to a given number of degrees of freedom. In this example there are three categories of flower colour, and there is the restriction that total deviation must be zero; this gives, $3 - 1 = 2$ degrees of freedom for χ^2. In general, if there are n categories, or columns in the table for calculating χ^2, there will be $(n - 1)$ degrees of freedom. In Table 80, for two degrees of freedom, when $p = 0.90$, $\chi^2 = 0.211$, and when $p = 0.80$, $\chi^2 = 0.446$. Thus in this experiment, for which it has been calculated that $\chi^2 = 0.234$ (Table 15)

Table 15

Observed and expected numbers of plants, and calculation of χ^2

	Red	Pink	White	Totals
Observed numbers (O)	145	289	138	572
Expected numbers (E)	143	286	143	572
Deviation ($O - E$)	2	3	-5	0
$(O - E)^2$	4	9	25	
$(O - E)^2/E$	0.028	0.031	0.175	$0.234 = \chi^2$

Flower colour spans Red, Pink, White.

p lies between 0.90 and 0.80. The interpretation of this is that one would expect deviations of this size in 90% to 80% of instances. To put it the other way round, there is a 10% to 20% chance that these results depart from the predicted $1:2:1$ ratio. Thus the chance of disagreement exceeds 5%, so we must reject the null hypothesis. Another genetical prediction must be made and tested.

Another application of the χ^2 test

Five traps were set in different locations within a wood; the numbers of field mice captured at each trap during a period of 3 months were recorded, and are given in the first row of Table 16. Upon capture, the mice were marked and released, so that they could be recognised again; there were few recaptures of the same mice, and these are discounted in the figures of Table 16. One trap (B) caught fewer mice than the others, and it was of interest to know whether this was merely a chance happening, or whether it was a significant difference, which might possibly be linked with the location of the trap.

Table 16

Observed and expected numbers of mice caught at traps, and calculation of χ^2

	Traps					
	A	B	C	D	E	*Totals*
Observed numbers (O)	23	7	25	19	21	95
Expected numbers (E)	19	19	19	19	19	95
Deviation ($O - E$)	4	-12	6	0	2	0
$(O - E)^2$	16	144	36	0	4	
$(O - E)^2/E$	0·842	7·579	1·894	0	0·211	$10·526 = \chi^2$

As in the previous example, one needs a null hypothesis to give a series of expected results, with which to compare the observations. In this example, however, there is no theoretical ratio which can be used, and since one is specially interested in discovering if B is trapping less mice than the others, a suitable null hypothesis is that there is *no difference* between the traps. If this is true, the expectation is that an equal number of mice will be caught at every trap. The total number caught was 95 mice so, for equality, there should be 19 caught at every trap. This figure has been entered in every column, A to E, in Table 16. The calculation proceeds as before, and it is found that $\chi^2 = 10·526$. There are four degrees of freedom (one fewer than the number of traps), From Table 80, χ^2 is 10·526 when p is between 0·05 and 0·02. If the null hypothesis is true, one would expect deviations as big as these in only 2% to 5% of instances. Thus, evidence is strongly *against* the null hypothesis, and it must be discarded. There is a 95–98% chance that the differences between the numbers caught at each trap really are significant. Inspection of Table 16 shows that the greatest contribution to χ^2 comes from trap B, so one may be confident that this trap is catching fewer mice than the others.

In these two examples, there were several categories to which the individual observations could be assigned. In the first example, there were three categories of flower colour, in the second, there were five categories of trap. The observations (excluding totals, expected numbers, and the rest of the calculation) were set out in a table with one row, and with a column for each category. A table like this, in which individuals may be assigned to one or other of the cells of the table is called a **contingency table**. The examples above were 1×3, and 1 × 5 contingency tables, since the tables are described by the number of **rows** and **columns** of cells they contain (again excluding columns for totals and rows for further calculations). The type of contingency table in these examples goes under the general description of a 1 × *n* contingency table. It has 1 row and *n* columns; whatever the value of *n*, it can be analysed by means of the techniques already described. Other contingency tables can have more than one row— these are called *n* × *n* tables. The simplest of these is the 2 × 2 table. This is commonly met with in biology, and the method of analysing it is given in the next section.

The 2 × 2 contingency table

Two batches of barley seed were to be tested to determine the effect of a certain heat treatment upon the viability of the seed. Batch A was the untreated, control batch, and Batch B was given the heat treatment. A sample of 80 seeds from each batch was then tested by the tetrazolium chloride method, to determine their viability. The seeds are split longitudinally with a sharp scalpel, immersed for half an hour in a 0·1% solution of 2,3,5-triphenyltetrazolium chloride, in the dark, and then examined. In viable seeds, the respiring embryo reduces the tetrazolium chloride to the intensely red, insoluble compound, triphenyl formazan. The intense coloration is easily distinguishable, and the number of viable seeds can be speedily determined. The results are set out in Table 17, where there are four cells or

The viability of 2 batches of seed, as a 2 × 2 contingency table.

Table 17 the observed results

	Viable	*Not viable*	*Totals*
Batch A	64	16	80
Batch B	34	46	80
Totals	98	62	160

Table 18 the expected results

	Viable	*Not viable*	*Totals*
Batch A	49	31	80
Batch B	49	31	80
Totals	98	62	160

Table 19 deviation from expectation

	Viable	Not viable	Totals
Batch A	15	− 15	0
Batch B	− 15	15	0
Totals	0	0	0

compartments, showing the numbers of viable and non-viable seeds in each batch. It is seen that Batch B contains fewer viable seeds than Batch A, or from the other point of view, that the viable seeds are mainly in Batch A, and the non-viable seeds are mainly in Batch B. Whichever way it is regarded, it seems that heat treatment reduces viability. Is this effect statistically significant? Could a table of figures like these have been obtained by randomly drawing two samples of 80 seeds from a large quantity of seeds, unaffected by heat treatment? If this had been done, the most likely result would have been that shown in Table 18. This preserves the column and row totals, but now shows no differences between the batches. Let the situation shown in Table 18 be the null hypothesis. Table 17 gives the observed results (O), and Table 18 gives the expected results (E), on the null hypothesis that there is no difference between the batches due to heat treatment. Next, a table is prepared, showing the deviations ($O − E$); as can be seen in Table 19, these deviations are equal, apart from their sign. This is a necessary consequence of keeping the row and column totals unchanged in arriving at Table 18. Note, at this point, how the procedure of calculation is following the same stages as those used in the earlier examples of the calculation of χ^2: this far, the values of ($O − E$) have been calculated, in Table 19. Since all the values of ($O − E$) are equal apart from sign, their squares will be equal and all positive; thus, ($O − E$)^2 is the same for every cell in the table, and the expression for χ^2 may be rewritten:

$$\chi^2 = \Sigma \frac{(O - E)^2}{E} = (O - E)^2 \cdot \Sigma \frac{1}{E}$$

In this example:

$$\chi^2 = (\pm 15)^2 \times (1/49 + 1/31 + 1/49 + 1/31) = 225 \times 0 \cdot 105\,34 = 23 \cdot 70$$

For a 2×2 contingency table, the number of degrees of freedom is always 1. There are four cells and three restrictions; (1) the grand total must be 160, (2) one row total must be 80, so fixing the other row

77

total at 80, and (3) one column total must be 98, so fixing the other column total at 62. This allows only one degree of freedom. By setting out a blank table, with the row, column, and grand totals filled in, it is possible to enter *any* figure in the first cell, but after that there is no further freedom. The various totals completely determine the other three entries.

In the table for χ^2, with one degree of freedom, a value of 23·70 corresponds to a probability considerably less than 0·001. If the null hypothesis is true, the probability of obtaining these results is less than 0·001. This strongly suggests that the null hypothesis is not true. If these two samples had been selected from a single batch one would obtain figures like these only once in 1000 experiments. Therefore, these batches cannot be considered as identical, or as part of the same batch. There is a significant difference between them; the effect of heat treatment upon the viability of the seed is significant at the 0·1 % level.

This example has been worked in a manner parallel to that used for the $1 \times n$ contingency tables of Tables 15 and 16. In practice, the value of χ^2 for a 2×2 contingency table may be calculated by a short-cut method, which is given on p. 177. This cuts out the intermediate stages of Tables 18 and 19, and allows one to go straight from the observed values to a value for χ^2.

The χ^2 test can be used with contingency tables of higher order, in which there are more than two categories for one or both of the two attributes being studied. For the procedure to be used, consult one of the texts listed on p. 205. Two further examples of the $1 \times n$ and 2×2 tables are given on pp. 110–17.

PROBLEMS

20 The barley variety *Albina 7* has a recessive mutant gene, which in the homozygote causes absence of chlorophyll. Plants with no chlorophyll cannot photosynthesise, and die shortly after germination. Seeds from self-pollinated heterozygous plants were germinated, and gave 12 white plants and 43 phenotypically normal green plants. Does this result support the genetical hypothesis, that the two types should appear in the ratio, *3 green : 1 white?*

21 In the F_1 generation of a cross between two strains of *Drosophila* there were 37 female flies, and 47 male flies. The expected sex ratio was, *1 male : 1 female*. Does the observed ratio differ significantly from expectation?

22 Two varieties of garden radish, one red and the other white, were crossed. The F_1 generation were all purple. These were allowed to interpollinate, and the F_2 seed collected. When germinated, this seed gave 42 red, 68 purple, and 50 white radishes. Theory predicts that the ratio of colours should be *1 red : 2 purple : 1 white*. Do these results support the theory?

23 Female *Drosophila*, homozygous for vestigial wings, were crossed with wild-type males. The F_1 generation all had normal wings, and were allowed to interbreed with one another. The F_2 generation consisted of 58 wild-type flies, and 13 flies with vestigial wings. Is this result in agreement with the expected ratio of *3 wild-type : 1 vestigial wings*?

24 Subjects were tested to see if they could, while blindfolded, distinguish between drops of distilled water, and drops of an infusion of cloves, placed on the tongue. In 500 tests the subjects were allowed to breathe normally (nose open), and in another 500 tests their noses were held closed while they attempted to tell which drops were on their tongues. Thus, in each test, the subject was asked to name the substance, and the verdict was recorded as correct or incorrect. The results of the 1 000 tests were:

	Verdict	
	Correct	*Incorrect*
With nose open	403	97
With nose closed	246	254

These figures appear to show that the subjects could more readily distinguish the substances when they were allowed to breathe freely, presumably because the olfactory cells of the nasal mucous membrane are the effective receptors of the smell of the substance placed on the tongue. However, this experiment is subject to considerable random error, owing to the low concentration of clove infusion employed, failure to wash out the mouth properly before the test, and many other reasons. Do the figures support the hypothesis that it is easier to distinguish the substances with the nose open, than with the nose closed?

9 Planning experiments

There is a tendency among biologists, especially among those who know little of statistics, to plan an experiment, perform it, work out the results, and only then to think about analysing the results statistically. This is a mistake, since failure to consider statistical aspects when the experiment is being planned may lead to a faultily designed experiment from which no valid or significant conclusions can be derived. On the other hand, an efficiently designed experiment may yield much information from relatively few experimental treatments, saving both time and resources. Statistical analysis is part of the technique of the experiment.

Controls

When planning a series of measurements to determine the effect of a given experimental treatment, one must also provide one or more series of control treatments with which the experimental treatment may be compared. For example, when experimenting on the effects of nitrogen deficiency on plants grown in water culture, one needs *experimental* plants, grown in a solution deficient in nitrogen, and *control* plants, grown in a complete mineral nutrient solution. This gives a basis for comparison; any differences between the two sets of plants may then be ascribed to the differences in their culture solutions, provided that *in all other ways* the culture conditions were identical for the two sets of plants. When the two sets of results (for example, the dry weights) have been obtained, they may be compared statistically, using the *t*-test to establish the significance of any difference between their means. Even if the *t*-test does indicate statistical significance, no biological significance can be attached to the results if the control treatment was not properly designed.

In some experiments, one control series can serve for several experimental series. The experiment of Table 1 is an example of this. In other experimental designs, a straightforward control treatment may not be needed; continuing the examples of experiments in mineral nutrition, one might study the effect of nitrogen level by growing plants in solutions containing 0·003M, 0·004M, 0·005M, and 0·006M potassium nitrate as the sole source of nitrogen, other mineral nutrients being supplied at constant level. Here there is no absolute control treatment, in the sense of complete absence of the factor being tested; one is not now interested in the effects of gross nitrogen deficiency. But there is still a logical basis for making comparisons—the solutions differ *only* in the level of nitrogen they supply—and these comparisons may be tested statistically. One might use the analysis of variance, or perhaps establish a correlation between nitrogen level and some aspect of growth, and calculate a regression equation.

Breeding experiments have no obvious control treatment, but again there is a comparison. One compares the observed ratios with those expected on the basis of what is believed to be the genetical mechanism underlying the experiment. Here one can judge the comparison by using the χ^2 test.

It is worth while at the planning stage of an experiment to give considerable thought to the designing of controls, or to making certain that the final results will be in a form in which valid and meaningful comparisons can be made. It is then important to decide upon an appropriate statistical technique with which to establish the significance of these comparisons, and to make sure that the experimental results will be adequate for the technique. Some other aspects of this last point will be considered in succeeding sections.

Precision of measurements

The data for the water content of turnips (p. 4) was obtained by weighing on a beam balance, which allows measurement to the nearest 0·1 g. The actual weights of the samples were about 100 g fresh and 10 g dry. Thus, the weights of the dry samples were determined to an accuracy of 0·1 g in 10 g, that is, to 1 %. This may seem to be a low order of accuracy, but from the figures on p. 4 one can see that samples from different turnips varied from one another by amounts appreciably more than 1 %. The variation due to differences between samples is greater than the inaccuracies due to experimental technique. As long as this is true, there is no point in using a more accurate method of weighing. A beam balance is far quicker to use than a chemical balance (except perhaps the very expensive automatic ones) and costs less to buy so that several can be available for use by a group of workers. Nothing would have been gained, and much time would have been lost, if one had chosen a chemical balance for weighing the turnip samples.

Points like this are worth bearing in mind when planning an experiment. Biological materials have inherent variability, and it is a waste of time to use measuring techniques which are of a considerably higher order of accuracy than the variability of the material. It may not always be possible to gauge the extent of variability until measurements have been completed and analysed, but experience will be a guide. When planning a very large experiment, in which there might be a considerable wastage of time and perhaps materials through the selection of too accurate measuring methods, it is a good idea to try a small pilot experiment first. This would be designed mainly to supply information on the extent of variability of the material under the various conditions it is proposed to use in the main experiment. The results of this experiment would indicate whether the measuring techniques were too accurate (or not accurate enough), would give some idea of a suitable number of replicates, and might uncover unexpected difficulties in the techniques or in the experimental design.

The number of replicates

As indicated in the last sentence of the preceding section, the number of replicates to be used may be determined by experience, or by trying a small-scale pilot experiment. With a small experiment, it is safer to err on the side of having too many replicates than too few. In a large experiment, too few replicates may lead to lack of significance in the results. Too many replicates may produce a highly significant result, but at the expense of extra work, extra materials, extra cost, and perhaps the sacrifice of information from treatments which had to be omitted on grounds of economy. It must be remembered also that no large number of replicates will enable one to prove an effect which does not exist.

The essential point about replicates is that they should, as far as is possible, be exactly alike. All differences between them may then be ascribed to random error, which contributes to the variance of the set of results. If there are *systematic* differences, perhaps relatively unimportant, these should be taken into account in the analysis. In Table 12, there was a systematic difference, not shown in the table, that for each treatment the four so-called replicates really comprised one leaf of four different area-grades. When this was taken into account (Table 13) and the analysis of variance modified, it was possible to extract the variance due to leaf area from the residual variance of the whole experiment. Though the effect of leaf area turned out to be barely significant, taking it into account in the analysis enables one to get a much better estimate of error (the residual variance) and to estimate the significance of the main treatment effect (jelly) more precisely.

In the example just referred to there were no replicates. An experiment *planned* on these lines can give much information in relation to the amount of effort involved in execution.

In breeding experiments, and in other studies where it is proposed to use the χ^2 test, planning must take account of the requirement that each cell of the contingency table must contain at least five observations. Thus, to test the significance of a $9:3:3:1$ ratio, there must be at least five individuals of the homozygous recessive genotype—an expected ratio of at *least* $45:15:15:5$. The experimenter must plan to obtain at least 80 individuals to obtain an analysable result.

Randomisation

The validity of many statistical techniques relies on the assumption that the differences between replicate individuals really are random. This assumption has a bearing on the technique used for selecting which individuals are to receive which treatment. Taking the simplest case, where the experiment is a straight comparison between a number of replicate individuals, receiving the experimental treatment, and a like number of individuals, receiving a control treatment, one must divide the individuals into two groups, choose one to be the experimental group and the other to be the control, and treat them accordingly. This apparently simple operation is a danger-point in experimental technique. Individuals vary, and it is easy, even if one is not *consciously* doing so, to pick out two groups which are not identical. At the extreme one may pick the healthiest-looking individuals for the experimental treatment, and use those which are left for the control. This violates the essential feature of a control; that it is identical with the experimental treatment *in all respects*, except the one factor under investigation.

Having realised the necessity of making experimental and control groups alike one may next attempt to pick out two groups which are as identical as can be arranged. It is tempting to try to do this, but can it be done? Perhaps one does not know enough to be certain that the two groups are identical. Having picked out a few for each group, and noticing that one has selected 'healthier' specimens for one of the groups one may try to compensate for this, by next picking weakly specimens for this group, in an attempt to equalise the groups. How does one know that one has succeeded in this aim? Whatever criteria are used to judge equality, how may one be certain that, perhaps unwittingly, some bias has been introduced into the experiment, a bias which destroys the validity of the result?

The only satisfactory solution to this problem is to avoid all wilful selection and rely instead on a random method of allocating individuals to treatments. A random method is one in which every individual has an equal chance of being allocated to every treatment. Even with random methods one need not get absolutely equal groups, indeed, it is *possible* for all the 'healthiest' specimens (or heaviest, or oldest, etc.) to be allocated to one treatment, but with the random selection procedure, the chances of this happening are small, and are calculable. More important, they are taken into account in the statistical analysis, when assessing the level of significance of the results.

The only way in which one can aim to secure equality between the groups is to begin with a uniform batch of plants or animals. This is not always easy to do, especially with the larger animals. To improve uniformity one may work through the batch, rejecting the obviously poor specimens and any others which are outstanding in any other way. This will give a uniform batch, from which one can then allocate individual members to the treatments, using one of the random methods which are described in the examples below. This rejection of extremes improves uniformity, and may give better defined experimental effects; on the other hand, one must not reject any members of the batch if the whole batch is to be regarded as a sample from a population. For instance, one cannot reject the antirrhinum plant with 165 flowers (Table 2) for it was a legitimate member of the sample and represented the more copiously-flowering members of the population. In that investigation there was no concern to secure uniformity, but to measure the statistical characteristics of the population.

Here are some examples of ways of randomising:
1. *Tossing coins*: Twenty insects are to be divided into four groups, each group of five insects to receive one of four different insecticide treatments, A, B, C, D. The insects are numbered from 1 to 20.

Taking each insect in turn, a coin is tossed twice. The scheme of allocation might be like this:

First toss	Second toss	Treatment
Head	Head	A
Head	Tail	B
Tail	Head	C
Tail	Tail	D

If, for example, the throws for insect No. 1 were 'Tail, head', that insect would be allocated to treatment C. The essential condition is that each of the four permutations of head and tail is equally likely to occur; for each of the permutations in the table, $p = 0.25$, so the condition for randomisation is satisfied. Towards the end, five insects will have been allotted to some of the treatments, and if a sixth insect obtains a throw which would put it into one of the completed groups, the coin must be thrown again. For the 20th insect no throw will be necessary.

In this way, the insects will have been allocated to their treatments so that every insect has had an equal chance of being selected for every treatment, provided that the coin itself is unbiased.

2. *Dice:* In the example above, an alternative method would be to throw a die and to decide beforehand that 1-spot means allocation to treatment A, 2-spots means B, 3-spots means C, and 4-spots means D. A throw which gives 5-spots or 6-spots is not counted.

3. *Playing cards:* Sixty insects are to be allocated to 12 treatments, five insects to each treatment. Take a pack of playing-cards and remove the Kings. For each insect in turn, shuffle the pack and cut it. The number of the card cut gives the treatment the insect is to receive, according to some prearranged scheme. For example, Ace is. treatment 1, 2 is treatment 2, and so on up to Queen for treatment 12. Playing cards are particularly useful when there is a large number of categories for allocation, but they can also be used for smaller numbers. In example 1, the four treatments could be indicated by the cutting of the four suits, spades, hearts, diamonds, and clubs.

4. *Telephone directory:* One hundred peas were sown in numbered pots (1–100) for a study on growth rate. Each week for 10 weeks, ten plants were to be uprooted for determination of dry weight and for chemical analysis. It would be very difficult to select ten plants each week without at least subconsciously picking perhaps the smallest or the largest ones, thus affecting the shape of the growth curve eventually obtained. The ten plants for harvesting on each occasion must be

selected at random, so that any of the over seventeen billion possible samples is equally likely to be chosen. It is possible to work out a method using cards, dice, or coins, but this example shows how to use a telephone directory. The book is opened at any page, and then the middle two figures from six-figure numbers, or the third and fourth figures from five-figure numbers are written down in the order that they appear on the page. For example, in my local directory, the numbers I write down are: 06, 89, 17, 55, 50, 45, 05, 26, 19, and 77. These would be the numbers of the plants chosen for harvest on the first occasion. The number 00 would be used to indicate plant No. 100. On the second and subsequent occasions, I would proceed down the page of the directory in this fashion, ignoring any repeats.

This method is a useful one, though there is always the possibility that there is some unsuspected bias; telephone numbers for certain areas are sometimes all prefixed by the same sequence of numbers.

5. *Table of random numbers :* Some workers have taken the trouble to prepare, once and for all, tables of numbers which are truly random in their occurrence. These published tables have been tested for randomness and contain no detectable bias. They are to be found in the *Cambridge Elementary Statistical Tables* and certain of the other books listed on p. 205.

Suppose that it is desired to sample the vegetation of an area of heathland, to put down quadrats at randomly chosen spots. One way of doing this is by having a circular or rectangular frame which is thrown down at random (over one's shoulder). The piece of ground marked out by the frame where it falls is the randomly chosen area. Repeated throws of this sort may be used to sample an extensive area, as in Problem 7 (p. 19). If the size of quadrat required is larger than the convenient size of a throwable frame, a stone or stick can be thrown, and a quadrat set up according to some pre-arranged scheme—say, the point of fall marks one corner, and two sides extend one metre to the north and to the west of this point, giving a one metre-square quadrat. The throwing method is simple to operate, but has a practical disadvantage that sooner or later the frame gets lost, or stuck up a tree. It has the theoretical disadvantage that there seems to be some bias introduced. The thrower tends to avoid walking in thick undergrowth, or to avoid throwing the frame into inaccessible places. Also if the frame falls on a tree, it may tend to be deflected away from the tree, and so fail to sample the vegetation adequately around the trunks of trees.

If a map of sufficiently large scale is available, the location of quadrats may be determined by drawing a system of coordinates on the map, and selecting points in this system by random numbers from tables, or using dice, cards or other methods. These randomly selected points are then located on the ground and sampling quadrats set up there. For example, starting at the south-west corner of the area, the coordinates might run from 0 to 60 metres in an eastward direction, and from 0 to 95 metres in a northward direction (metres on the ground or scale metres on the map, it does not matter). Stabbing with a pointer, without looking, at the open page of a table of random numbers, the pointer first indicates '24 28'; let this indicate 24 metres east, 28 metres north. The next pair of numbers below these is 42 43 (42 m east, 23 m north), then 08 12 (8 m east, 12 m north) and so on. Having marked these points on the map, they can then be located on the ground and sampled. The subject of sampling in ecological work is given full consideration by Greig-Smith (1957).

Latin squares

Suppose that an experiment is being designed to investigate the effect of four fertiliser treatments on the growth of a crop plant. A plot of land is to be divided into four equal areas, each area to receive one of the four treatments before the seed is sown. In what way is the area to be divided? It could be divided into quarters, as in Fig. 12a. If the

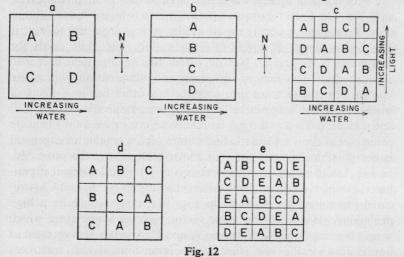

Fig. 12

slope of the land was such that the soil to the east was moister than that to the west, plants receiving fertiliser B and D would also receive more water. The effect of the water might even be greater than the effect of the fertilisers and so ruin the experiment. Even a small effect of water would bias the comparison between A or C and B or D.

The bias might be minimised by dividing the land into strips, as in Fig. 12b. Now each treatment extends over both drier and moister soil. If there are trees to the south of the field, D will be in the shade for more of the day, so another bias will be introduced. This may be an extreme example, but it illustrates how environmental factors may introduce bias into an experiment. Even if care is taken to provide as uniform an environment as possible, by improved drainage, or by felling the trees to the south of the field, one still cannot be sure that other, unrecognised factors are not upsetting the experiment— the direction of the prevailing wind, previous fertiliser treatments, and other less clearly defined influences. In a greenhouse one may meet similar variations; illumination may be uneven in a complex pattern, watering may be less copious at the ends of the bench than in the middle, and less at the back than at the front, and temperature may be lower nearer the glass than nearer the centre of the greenhouse. There are so many environmental factors, operating along so many gradients, that it is impossible to allow for them all. The latin square is one way of nullifying their effects.

To form a latin square, the field is divided into 16 smaller squares, as in Fig. 12c. Each treatment appears in one row and in one column only. Treatment A occurs on all grades of soil, from dry to wet; it also occurs under all levels of illumination. The same holds for treatments B, C, and D. Not only does this arrangement take into account those two factors (drainage and illumination) which are known, but it also takes into account any other factors (effects of wind, etc.) which may not be known, or even suspected. Figures 12d and 12e show 3×3 and 5×5 latin squares constructed on the same principles as the 4×4 square. Sometimes such a regular arrangement is not possible, but one can still contrive to achieve the same end. In Fig. 13a there are 15 culture vessels of a mineral nutrient experiment, in which there are three different solutions (A, B, and C) replicated five times. Arranged as in Fig. 13a, there would be a high probability of there being some factor such as temperature which would bias the difference between A and C. A better arrangement of pots is shown in Fig. 13b. Here, each column contains each treatment

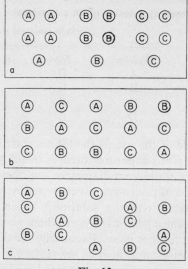

Fig. 13

once only, and each row contains each treatment no more than twice. Even so, the arrangement is not perfect. Another arrangement would be as in Fig. 13c, which uses the 5×5 latin square from Fig. 12e, leaving spaces where D and E occur.

So far, we have considered a latin square to be a set of plots laid out on the ground, or an arrangement of culture vessels on a bench. The same idea can be extended to the *design* of experiments, even when a physical layout of plots or vessels is not involved. For example, the experiment with Cherry Laurel leaves (Chapter 6) required four treatments with jelly, and leaves of four grades of size, so that 16 leaves were needed for the experiment (Table 13, p. 61). We could have made the experiment more complex by choosing the leaves from four different varieties of Cherry Laurel, provided that the leaves of different varieties were assigned to their treatment and selected from their size grades in a regular way. The basis for the design is a 4×4 latin square. One such square, chosen at random from the possible 4×4 latin squares, is shown in Table 20. The four varieties are represented by the four letters, to each of which any one variety would be allocated at random. By using this design each variety is subjected once to each treatment, and one leaf from each grade is used. Although the design still uses only 16 leaves, we now have information about three factors—treatment, size, and variety—instead of only two factors.

D

Table 20

A 4 × 4 latin square used as a design of an experiment

Area of leaf	Surface covered with jelly			
	Neither	*Top*	*Bottom*	*Both*
Grade 1 (smallest)	A	D	B	C
2	B	C	A	D
3	C	B	D	A
4 (largest)	D	A	C	B

Another example of the use of the latin square design is given in Table 61, p. 170. In this experiment leaves of tulip were punched to make circular discs of different diameters, and these discs were made to sink in a solution of potassium bicarbonate by replacing the air in the mesophyll spaces with the solution.[1] When the discs photosynthesise they produce gaseous oxygen which accumulates in the air spaces, perhaps with other gases. Eventually the discs are buoyed by the accumulating gases, and float to the surface of the solution. The time taken for this to occur can be measured, and thus this technique is a simple though crude way of assessing the effect of various experimental conditions on the rate of photosynthesis. In the experiment for which data are given, two experimental factors were being tested: *Light intensity* at four levels, 1 to 4, corresponding to 80%, 50%, 13% and 3% of full sunlight, respectively; *temperature* at four levels, 8°, 14°, 21° and 28° C. It was decided to see what effect leaf diameter had upon the times taken to rise to the surface, so discs were punched in four sizes, their diameters being, A = 5 mm, B = 4 mm, C = 3 mm, and D = 2 mm. Here we have a 4 × 4 latin square design, with three factors being tested. The data is examined by an analysis of variance, according to the method described on pp. 170–2. This method is very similar to the method previously considered. For the effect of light the variance ratio is $F = 4/3·33 = 1·2$, and this is not statistically significant. Tulip grows perfectly well in shady situations where light intensity is often as low as 3% of full sunlight, so this is not surprising. For the effect of temperature, we find $F = 185/3·33 = 55·5$. This is highly significant ($p < 0·001$), the rate of photosynthesis appearing to increase with increasing temperature, within the range under examination. For the effect of disc size we find $F = 51/3·33 = 15·7$, for which p lies between 0·01 and 0·001. This is a significant effect which has a bearing on the technique of the

[1] See Bishop, O. N. (1971), *Outdoor Biology, Teachers' Guide*, Murray.

experiment. One possible explanation is that the punching machine causes damage to cells lying within a fixed distance of the cutting edge, so that when the diameter is smaller a proportionately greater number of cells lie within this distance. Thus the smaller discs contain a smaller proportion of actively photosynthesising cells.

The latin square design offers the advantage of giving more information from the same number of observations, but it has two important features which may sometimes be disadvantageous. The first is that the number of levels of each factor *must* be equal. For example, we must have four light levels, four temperatures, and four sizes of disc. It is not possible to study five levels of temperature without increasing the number of light levels and disc sizes to five. The only solution is to adopt a modification of the latin square, called a Youden square; this design is dealt with in the more advanced texts. The second feature is that the latin square can be used only if the factors show no **interaction**. An interaction occurs if the effect of one factor is partly dependent upon one or more of the other factors. If, for instance, at lower temperatures the rate of photosynthesis were to increase with increasing light intensity, but at higher temperatures it was unaffected by light intensity, or showed a *de*crease with increasing light intensity, we would say that there was an interaction between the effects of temperature and the effects of light. There is no reason to suppose that at the levels used in this investigation there will be any interaction between the three factors, so the use of the latin square is justified. If interaction is suspected, and we wish to test for its existence we must use a factorial design, as explained in the next section.

Before leaving the subject of latin squares, there is one special type which may be useful in certain investigations. We might wish to investigate the effect upon female African clawed toads of various levels of injected gonadotrophic hormone at various times during the year. The three factors would then be: hormone dosage, date of injection, and differences between individual females (for these might be from different sources, and of different ages). We could use six levels of dosage, six dates, and use six female toads. At each injection date the six females would receive one of the six levels of injection, according to the design of the 6 × 6 latin square chosen for the experiment. The difficulty would be that the effect of an injection might depend partly on the levels of injections previously given to that individual, particularly the injection immediately before the one then under test. To obviate this difficulty we use a special type of latin square, constructed as follows:

If the square is of *even* order $u \times u$ (4×4, 6×6 etc), we write down the first row of the table as a series of numbers:

$$1 \quad 2 \quad u \quad 3 \quad u-1 \quad 4 \quad \text{and so on.}$$

The sequence 1, u, $u-1$, $u-2$, etc, alternates with the sequence 2, 3, 4, etc. For example, for a 6×6 square the sequence would be: 1, 2, 6, 3, 5, 4. Then we write the second row of the table by adding 1 to each number in the first row, subtracting 6 from any number over 6. This is repeated for the third and following rows. This gives us a table:

1	2	6	3	5	4			A	B	F	C	E	D

1	2	6	3	5	4
2	3	1	4	6	5
3	4	2	5	1	6
4	5	3	6	2	1
5	6	4	1	3	2
6	1	5	2	4	3

which can be written as letters, if preferred.

A	B	F	C	E	D
B	C	A	D	F	E
C	D	B	E	A	F
D	E	C	F	B	A
E	F	D	A	C	B
F	A	E	B	D	C

To use this table, the rows would represent the individual toads, the columns the treatment dates, and the figures or letters the different levels of hormone. Examination of this square shows that each hormone level follows each other hormone level only once.

If the square is of *odd* order (5×5, 7×7 etc), we need to write out two squares. The first row of the first square is:

$$1 \quad 2 \quad u \quad 3 \quad u-1 \quad 4 \quad \text{etc}$$

The first row of the second square is the reverse of the first row of the first square. Combining two such 5×5 squares we get:

1	2	5	3	4	⎫
2	3	1	4	5	⎪
3	4	2	5	1	⎬ first square
4	5	3	1	2	⎪
5	1	4	2	3	⎭
4	3	5	2	1	⎫
5	4	1	3	2	⎪
1	5	2	4	3	⎬ second square
2	1	3	5	4	⎪
3	2	4	1	5	⎭

In an experiment with hormone treatment on toads, we would test five levels of hormone, on five dates, and would need 10 toads. Each toad would receive each level of hormone once during the course of the experiment. Each level follows each other level twice.

Interaction

An explanation of interaction was given on p. 91. Interactions occur fairly commonly and their existence can be tested for in the analysis of variance. The essentials of the experimental design are that there should be at least two experimental factors, and that each combination of treatments should be replicated. The minimum requirements are two factors, applied at two levels, with two replicates of each combination of the two factors. In more complex designs as many as five factors can be tested, each applied at several levels, and the interactions between pairs or trios or quartets of these factors—or all five together—can be tested for. The technique of analysis is complex, and the reader is referred to a more advanced text. Here we will discuss the analysis of a two-factor experiment, with replication. The data of such an experiment is presented in Table 64 (p. 172). The experimental method was the same as that described for the latin square design in the previous section, but there are differences:

1. Only two factors are tested—light intensity and temperature, each at four levels (though there is no *need* in this design for the number of levels for one factor to be the same as the number of levels of the other).
2. The leaf discs are 5 mm in diameter for all treatments.
3. Each treatment is replicated twice.
4. The total number of observations is 32.

By comparison with a latin square, we have more observations to make, and test only two factors, but we are able to gain information on interaction. The analysis of variance (Table 65, p. 173) gives $F = 4/3 \cdot 1 = 1 \cdot 3$, for the effect of light, so this shows no significant effect (as in the latin square experiment). For the effect of temperature we find that $F = 167/3 \cdot 1 = 54$, which is highly significant ($p < 0 \cdot 001$). For the interaction between light and temperature we find $F = 3 \cdot 8/3 \cdot 1 = 1 \cdot 2$. Interaction is not significant and this, incidentally, justifies the use of the latin square design, as far as the light × temperature interaction is concerned. By this analysis we have been able to proceed further than a two-factor analysis without replication; we have been able to partition the residual mean-square of such a two-factor analysis into two fractions, one of which represents the interaction and the other of which represents the true residual error. It is only by having replicates that we are able to do this. The two-factor analysis *without* replication (pp. 166–70) contains, in its residual mean square, a mean square for interaction (if interaction exists). In the example given, the residual mean square is small and the main treatment effect is significant, so we can afford to ignore the possibility of interaction. Had the residual mean been larger, and no significant treatment effect been demonstrated, we might suspect that interaction was inflating the value of the residual mean square. The only course would have been to repeat the experiment, with replication, so as to estimate the effects of interaction.

In the analysis of Table 65 (p. 173), no interaction is shown but, had there been an effect, the next step would be to break down the data into separate tables for each level of light intensity, and to analyse each table to discover the effects of temperature. It might be found that at one light level, the temperature had a significant effect, while at another it had no effect. Thus we could unravel the nature of the interaction. One could also break down the data by temperature levels, and test the effect of light level at each temperature, separately.

10 Statistics in action

In this chapter are some examples of the types of biological problem which can be tackled by using the statistical techniques already described. No fundamentally new techniques are used for these examples, though slight variations of method are used, where the data require or permit it.

1 A comparison of two phases of a bracken community

A bracken community comprised two phases, one an active 'building' phase, and the other a 'degenerate' phase. It was decided to estimate and compare the production of fronds in each phase. Accordingly, the phases were sampled by placing 30 quadrats, each measuring 0.3 m $\times 0.3$ m, within each phase. The number of fronds found in each quadrat was counted, and is referred to as the frond density. The results, as recorded, are given in Table 21. There are two

Table 21

Frond densities in two phases of a bracken community

	The number of fronds found in 0.3 m $\times$ 0.3 m quadrats									
Building phase	9	8	7	6	7	8	7	9	3	9
	0	0	2	6	14	5	8	19	0	7
	18	3	9	2	1	1	16	13	8	11
Degenerate phase	5	4	0	1	0	2	16	3	0	7
	3	1	0	0	11	0	2	3	19	14
	0	3	0	0	1	4	6	8	13	0

samples to compare. One has to decide whether they come from a single population, or from two different populations. At first glance at the figures in Table 21, it is not possible to decide very much, since the data are arranged so haphazardly. When the figures have been arranged as frequency distributions, and when the histograms of these distributions have been drawn (Fig. 14), it begins to look as if the frond density is greater in the building phase. In this phase, the mode is at 7, 8, and 9 fronds per quadrat, while in the degenerate

Statistics for biology

phase the mode is at zero. However, in both phases there are a few quadrats with high numbers of fronds, and one must obtain more critical evidence before passing judgement. The appropriate test is to calculate the means and variances of each sample, and then use the *t*-test to decide whether the means are significantly different. This is the same procedure as was used for the data on the flower number of the two varieties of antirrhinum.

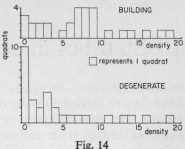

Fig. 14

The calculation of means and variances is shown in Tables 22 and 23, where the short-cut methods and checks on correctness have been employed, as described on pp. 155–9. Readers who have not

Table 22

Frond density: calculation of mean and variance

THE BUILDING PHASE:

Frond density Frequency

x	f	D	fD	$D+1$	$f(D+1)$	fD^2	$f(D+1)^2$
0	3	-7	-21	-6	-18	147	108
1	2	-6	-12	-5	-10	72	50
2	2	-5	-10	-4	-8	50	32
3	2	-4	-8	-3	-6	32	18
4	0	-3	0	-2	0	0	0
5	1	-2	-2	-1	-1	4	1
6	2	-1	-2	0	0	2	0
7	4	0	0	$+1$	4	0	4
8	4	$+1$	4	$+2$	8	4	16
9	4	$+2$	8	$+3$	12	16	36
10	0	$+3$	0	$+4$	0	0	0
11	1	$+4$	4	$+5$	5	16	25
12	0	$+5$	0	$+6$	0	0	0
13	1	$+6$	6	$+7$	7	36	49
14	1	$+7$	7	$+8$	8	49	64
15	0	$+8$	0	$+9$	0	0	0
16	1	$+9$	9	$+10$	10	81	100
17	0	$+10$	0	$+11$	0	0	0
18	1	$+11$	11	$+12$	12	121	144
19	1	$+12$	12	$+13$	13	144	169
Totals	30		$61-55=6$			774	816

Working mean, $w=7$ Class interval, $c=1$.

CHECK: $816 - 774 - (2 \times 6) = 30$; working correct so far.

$$\bar{x} = 7 + \left(1 \times \frac{6}{30}\right) = 7 + 0\cdot2 = 7\cdot2 \text{ fronds}$$

$$\Sigma d^2 = 1^2 \left(774 - \frac{6^2}{30}\right) = 774 - 1\cdot2 = 772\cdot8$$

$$\sigma^2 = \frac{772\cdot8}{30-1} = 26\cdot65$$

Table 23

Frond density: calculation of mean and variance

THE DEGENERATE PHASE:

Frond density x	*Frequency* f	D	fD	$D+1$	$f(D+1)$	fD^2	$f(D+1)^2$
0	10	− 5	− 50	− 4	−40	250	160
1	3	− 4	− 12	− 3	− 9	48	27
2	2	− 3	− 6	− 2	− 4	18	8
3	4	− 2	− 8	− 1	− 4	16	4
4	2	− 1	− 2	0	0	2	0
5	1	0	0	+ 1	1	0	1
6	1	+ 1	1	+ 2	2	1	4
7	1	+ 2	2	+ 3	3	4	9
8	1	+ 3	3	+ 4	4	9	16
9	0	+ 4	0	+ 5	0	0	0
10	0	+ 5	0	+ 6	0	0	0
11	1	+ 6	6	+ 7	7	36	49
12	0	+ 7	0	+ 8	0	0	0
13	1	+ 8	8	+ 9	9	64	81
14	1	+ 9	9	+ 10	10	81	100
15	0	+ 10	0	+ 11	0	0	0
16	1	+ 11	11	+ 12	12	121	144
17	0	+ 12	0	+ 13	0	0	0
18	0	+ 13	0	+ 14	0	0	0
19	1	+ 14	14	+ 15	15	196	225
Totals	30		$54 - 78 = - 24$			846	828

Working mean, $w = 5$ Class interval, $c = 1$.

CHECK: $828 - 846 - (2 \times - 24) = 30$; working correct so far.

$$\bar{x} = 5 + \left(1 \times \frac{-24}{30}\right) = 5 - 0\cdot8 = 4\cdot2 \text{ fronds}$$

$$\Sigma d^2 = 1^2 \left(846 - \frac{(-24)^2}{30}\right) = 846 - \frac{576}{30} = 846 - 19\cdot3 = 826\cdot7$$

$$\sigma^2 = \frac{826\cdot7}{30-1} = 28\cdot51$$

yet used this technique need not be concerned with its details at present; it is given here as an additional worked example.

The means, 7·2 and 4·2 fronds, have a relatively large difference, which appears to support the initial suggestion, that the quadrats in the building phase have a higher frond density than those in the degenerate phase. The next step is to find the significance of this difference, using the *t*-test, as described in Chapter 5. In the calculations which follow, the suffix $_1$ refers to the building phase, and the suffix $_2$ refers to the degenerate phase.

The means to be compared are: $x_1 = 7·2$, and $x_2 = 4·2$

Their variances are: $\sigma_1^2 = 26·65$, and $\sigma_2^2 = 28·51$

Both samples contained 30 quadrats, so: $n_1 = n_2 = 30$

The variance of the difference of means, σ_d^2, is given by:

$$\sigma_d^2 = \frac{\sigma_1^2}{n_1} + \frac{\sigma_2^2}{n_2} = \frac{26·65}{30} + \frac{28·51}{30} = 0·8883 + 0·9503 = 1·8368$$

The standard deviation of the difference between means is:

$$\sigma_d = \sqrt{1·8386} = 1·356$$

$$\text{and} \quad t = \frac{7·2 - 4·2}{1·356} = 2·21$$

Putting this result into words, the difference between means is 2·21 times its standard deviation. From the table of *t* (Table 74), for $p = 0·05$, and 60 degrees of freedom ($n_1 + n_2 - 2 = 58$; nearest is 60), the value of *t* is 2·00. The difference of means is therefore significant at the 5% level.

There is one important reservation which must be made about the result of this analysis. The calculations and tests of significance are all based on the assumption that the data conform to the *normal distribution*. The histograms shown in Fig. 14 present us with no evidence that this is so for either set of data. Maybe, if more data were to be collected, the histogram of the building phase would eventually approximate to a normal distribution, but the histogram for the degenerate phase would probably follow the already apparent trend of having its mode at zero, with frequency falling sharply away at higher frond densities. There are ecological reasons for believing that this may be so, and that the distribution in the degenerate phase, and perhaps in the building phase too may not be normal. We call an asymmetrical distribution with one mode a **skew distribution**.

Special statistical methods, not described in this book, are available for dealing with skew distributions. In most examples that one meets the amount of skewness is moderate, and no serious error will arise if the distribution is treated as a normal one, following the methods of this book. In the example dealt with here, it looks as if the skewness is very marked, so that the assessment of the level of significance of difference of the means is not reliable. Before admitting that a statistically significant result had been obtained one would insist on a higher value for t, say around 2·66 ($p = 0·01$). Thus, with a low value of t and the likelihood of strongly skew distributions, this data is not convincing proof that there is a significant difference between the frond densities in two phases.

2 *A comparison of the lengths of chromosomes*

By special techniques it is possible to culture the white cells from human blood. Stained preparations made from these cultures show some of the cells in mitosis, when the chromosomes can be clearly distinguished. Many of the chromosomes of a cell can be paired and identified by virtue of their length and the position of the centromere. For example, the longest pair is conventionally identified as pair No. 1. The next longest pair is known as pair No. 2, and so on. Pairs No. 6 to 12 are very similar in size, and it is not at present possible to pair any two chromosomes and state with confidence that they are a homologous pair. It is best to regard these chromosomes as a group, and to include within this group the X chromosomes, which are also of similar length. Some workers maintain that the X chromosomes (or the single X chromosome in the cells of a male) are the longest in the 6–12 group, while others have suggested that the X chromosome occupies one or another stated position within that group. By a special technique of treating the cells with a culture solution containing tritiated thymidine, it is possible to induce one of the X chromosomes of female cells to take up the radioactive substance while the other chromosomes do not. Later, when a thin film of photographic emulsion is placed over a microscopical preparation from the culture, and developed, this X chromosome is revealed because the film in its vicinity is exposed to radiation, which causes the formation of an image of silver grains. Thus it is possible to identify one of the X chromosomes and to distinguish it from other chromosomes in the 6–12 + X group. This technique can be applied only to female blood cultures, and it allows the identification of only one of the two X chromosomes in each cell.

Statistics for biology

Using the technique described above, preparations were obtained which were photographed. The enlarged photomicrographs of 35 cells, five cells from each of seven different females, were printed to a known degree of magnification. In these photomicrographs the lengths of the 16 chromosomes in the $6-12+X$ group were measured,

Table 24

The lengths of chromosomes

(All lengths are in micrometres, rounded off to the nearest $0\cdot1$ μm.)

Female No.	Cell No.	Length of X (x)	Length of 'the rest' (y)	Rest − X (z)
1	1	4·5	4·2	−0·3
	2	5·4	5·0	−0·4
	3	4·4	4·4	0
	4	6·1	5·9	−0·2
	5	4·6	5·2	0·6
2	1	5·0	5·4	0·4
	2	6·3	6·8	0·5
	3	8·2	7·1	−1·1
	4	7·8	6·3	−1·5
	5	6·6	5·9	−0·7
3	1	2·4	2·7	0·3
	2	3·5	3·5	0
	3	4·6	4·8	0·2
	4	5·0	5·8	0·8
	5	3·8	3·4	−0·4
4	1	3·5	3·4	−0·1
	2	3·9	4·4	0·5
	3	5·3	4·4	−0·9
	4	8·3	7·1	−1·2
	5	7·3	7·5	0·2
5	1	4·2	4·4	0·2
	2	5·1	4·5	−0·6
	3	3·6	3·7	0·1
	4	5·7	6·2	0·5
	5	4·9	4·6	−0·3
6	1	5·1	5·4	0·3
	2	5·7	5·7	0
	3	4·4	4·2	−0·2
	4	4·9	4·6	−0·3
	5	5·9	6·0	0·1
7	1	4·1	4·0	−0·1
	2	4·7	4·6	−0·1
	3	4·2	4·1	−0·1
	4	4·5	4·0	−0·5
	5	4·5	4·0	−0·5

and by multiplication by a magnification factor, their absolute lengths were calculated. The length of the identifiable X in each cell is given in Table 24. Alongside it is given the mean length of the other chromosomes of this group in the same cell. Since these 15 other chromosomes (pairs Nos. 6 to 12, plus the *other* X chromosome) are not distinguishable from one another, it has been found more convenient to treat them as a unit, and to compare them with the one identifiable X chromosome. This will from now on be referred to simply as 'the X chromosome', and the rest of the 6–12 + X group will be referred to as 'the rest'.

From Table 24 it is apparent that there is considerable variation in chromosome size from cell to cell. This may be due to several factors, such as the exact stage of mitosis reached by the cell immediately prior to fixation, and the degree of osmotic swelling of the cell and its contents during the preparation of the microscope slides. Variations of this kind seem to affect all the chromosomes in a cell similarly. Ignoring this variation in size, what is to be decided is whether the X chromosome is significantly greater or smaller than the mean of 'the rest'. This will give an indication of its true position within the group.

If the simplest approach to the analysis were to be taken, one would adopt the same technique as that used in Example 1 (p. 95). The 35 X chromosomes would be treated as a single sample, and the means of 'the rest' as another sample. The means and variances of these two samples would be calculated, and the final stage would be to try to establish the significance of the difference between the means. This approach is not the best one available for the present data, because it does not make use of important additional information which the data provide. For each cell the length of the X chromosome has been measured, and *for the same cell*, the mean length of 'the rest'. To amalgamate the data for the X chromosomes, as would be done when summing their lengths to calculate the mean, and then to do the same with the mean lengths of 'the rest', is to throw away valuable information. In this problem it is better to compare the length of each X chromosome with the mean length of the other chromosomes in the group, found in the same cell. The best quantity to work with is the *difference* between the length of the X and the mean length of 'the rest'. This difference is given in the column on the right of Table 24, and is referred to as z.

Examination of these figures shows that there is little consistency in the size of X relative to that of the mean of 'the rest'. Of the 35

101

values of z, 13 are positive and 19 negative, suggesting that there is no consistent difference. Indeed, a reasonable hypothesis is that there is no significant difference between x and y. This is the null hypothesis of the analysis—that the mean z is not significantly different from zero. To test this one must calculate $\bar{z}$, and its limits. The later stages of calculation are:

$$\Sigma z = -4\cdot8 \qquad \Sigma z^2 = 10\cdot00 \qquad n = 35$$

$$\therefore \; \Sigma d^2 = 10\cdot00 - \frac{(-4\cdot8)^2}{35} = 10\cdot00 - 6\cdot583 = 3\cdot417$$

$$\sigma^2 = \frac{3\cdot417}{35-1} = 0\cdot1005$$

$$\sigma = \sqrt{0\cdot1005} = 0\cdot3170 \qquad \sigma_n = \frac{0\cdot317}{\sqrt{35}} = 0\cdot05358$$

By interpolation in the table of t, for 34 degrees of freedom, and for $p = 0\cdot05$, we find $t = 2\cdot03$. The limits are $\pm 0\cdot05358 \times 2\cdot03 = \pm 0\cdot109$

$$\bar{z} = -4\cdot8/35 = -0\cdot137$$

$\therefore$ The mean difference between the X and the mean of the rest,

$$\bar{z} = -0\cdot137 \pm 0\cdot109 \text{ micrometres.}$$

The length of the X chromosome is thus significantly greater than the mean of the rest of the group, there being a 95 % chance that z has a value between $-0\cdot246$ and $-0\cdot028$. One can now state that the X definitely belongs to that half of the group containing the longest members, though to make a statement more precise than this, and to allocate it to any definite position within the group, would require considerably more measurements than there are here—in fact, it may prove impossible to resolve this problem of identification of these chromosomes solely on the basis of length.

Taking $p = 0\cdot01$, the table gives $t = 2\cdot73$, making the limits of the mean, $\bar{z}$, increase to $\pm 0\cdot146$. At the 1 % level of significance one can say only that the mean lies between $-0\cdot283$ and $+0\cdot057$, a range which includes zero. At the 1 % level of significance, therefore, one cannot exclude the null hypothesis that the X is not different in length from the mean of the rest.

Before leaving this example, it is interesting to see what would have happened with the method of the previous example. Calculations for the two samples would have given:

$$\bar{x} = 5\cdot086 \text{ micrometres}$$
$$\bar{y} = 4\cdot949 \text{ micrometres}$$
$$\bar{x} - \bar{y} = 5\cdot086 - 4\cdot949 = 0\cdot137 \; (= -\bar{z})$$
$$\sigma_d = 0\cdot955$$

$$\text{Thus } t = \frac{0 \cdot 137}{0 \cdot 955} = 0 \cdot 143$$

The differences of the means is the same as the mean of the differences previously calculated (excepting for the sign, because the difference is taken the other way about) and to be significant at the 5 % level, t must be greater than the tabulated value. For 68 degrees of freedom, and for $p = 0 \cdot 05$, $t = 2 \cdot 00$. The calculated value thus is much less than the tabulated value, and the analysis gives no evidence of a significant difference between the length of the X and the mean length of the rest. This illustrates how important it is to retain all available information in the calculation by using the method first recommended for this example.

This technique may be applied to any two series of data, for which the variable has the same units, and for which there is a legitimate reason for pairing every observation of one set with a specified observation from the other set. For instance, if a nutritionist has a dozen rabbits, and they are each weighed before and after a course of a special diet, he can pair the data for each rabbit ('weight before' with 'weight after') to see whether there has been any consistent change of weight during the period of treatment. There may be quite large differences between the individual rabbits, just as there were large intercellular differences in this example, but the point of interest is the effect of diet on the weight of each rabbit. By calculating the differences in weight before and after treatment, for each rabbit, and by working with this derived data, the nutritionist can extract the required information.

3 *A closer look at the effects of mineral deficiency*

The data used for this example are the same as those given in Table 1, except that an extra treatment (lacking calcium) has been included. The data are set out afresh in Table 25. For the moment, ignore the columns headed 'standard deviation' and 'S.D./mean', and pass on to Table 26, where the stages in the calculation of sums of squares are set out, using the technique recommended on pp. 184–7, for use with a hand calculating machine.

The analysis of variance of the data is set out in Table 27. There are 15 observations, with one restriction (that $\Sigma d = 0$), so there are 14 degrees of freedom. Of these, four belong to the differences between treatments (five culture solutions, one restriction, that the total of all treatments shall be 4659), and the remaining ten degrees of freedom are assigned to residual variation. For the comparison of between-treatments mean square with residual mean square,

$$F = 342211/12522 \cdot 6 = 27 \cdot 3.$$

For $n_1 = 4$, $n_2 = 10$, and $p = 0 \cdot 001$, Table 78 gives $F = 11 \cdot 3$. The effect of treatment is therefore highly significant.

Table 25

Data of Table 1, analysed further

| Culture solution | Dry weight (mg) | | | | Standard | |
	Plant 1	Plant 2	Plant 3	Mean	deviation	S.D./mean
Complete	1172	750	784	902	191	0·21
Lacking calcium	108	173	174	152	31	0·20
Lacking magnesium	67	95	59	74	15	0·20
Lacking nitrogen	148	234	92	158	58	0·37
Lacking micro-nutrients	297	243	263	268	22	0·08

Table 26

Calculation of sums of squares of data of Table 25
(dry weights in mg)

| Culture solution | Replicates | | | | Sums of squares |
	1	2	3	Totals	
Complete	1172	750	784	2706	2550740
− Ca	108	173	174	455	71869
− Mg	67	95	59	221	16995
− N	148	234	92	474	85124
− micro-nut	297	243	263	803	216427
			Totals	$G = 4659$	$H = 2941155$

$K = 8447787$ $\qquad K/3 = 2815929$

$\qquad\qquad\qquad\qquad G^2/15 = 1447085$

$u = 3$ $\qquad\qquad v = 5$

Total SOS $\qquad\quad = 2941155 - 1447085 = 1494070$

'Between rows' SOS $= 2815929 - 1447085 = 1368844$

Residual SOS $\qquad = 1494070 - 1368844 = 125226$

Table 27

Analysis of variance of data from Table 25

Source of variance	Sum of squares	Degrees of freedom	Mean square
Treatments between rows	1368844	4	342211
Residual	125226	10	12522·6
Totals	1494070	14	

This is very encouraging, but what is implied by this highly significant result? Does it mean that *all* treatments have significantly differing means? Or are only one or two of the means different?

To compare any two treatment means, one *should* calculate their variances and perform the usual *t*-test, so that to be accurate and to compare all possible pairs of means, the variances for each one of the five treatments should be calculated separately. However, as a rough approximation, one can use the residual mean square, and consider that the variance is the same for all treatments. A more exact technique for estimating the significance of the differences of means of more than two samples is given by Snedecor (1956), p. 251.

Applying the method of p. 48 to the difference between any two means of three observations each we have:

$$\sigma_1^2 = \sigma_2^2 = 12\,522\cdot6 \text{ (the residual mean square)}$$

and $n_1 = n_2 = 3$

so $\sigma_d = \sqrt{\left(2 \times \dfrac{12\,522\cdot6}{3}\right)} = \sqrt{(8\,348\cdot4)} = 91\cdot37$

To make the difference of means significant, the value of *t* must exceed the value taken from the table. The residual mean square has ten degrees of freedom, and for $p = 0\cdot05$, Table 74 gives $t = 2\cdot23$. To be significant, a difference between means must exceed $2\cdot23$ times its standard deviation. Thus the difference between two means must exceed $2\cdot23 \times 91\cdot37 = 204$ mg.

Table 28

Differences of means (mg), and their significance

Greater mean	Lesser mean			
	− Mg	− Ca	− N	− micro-nutrients
Complete	728***	750***	744***	634***
− micro-nutrients	194	116	110	—
− N	84	6	—	—
− Ca	78	—	—	—

*** significantly different ($p = 0\cdot001$).
Other differences are not significant.

The differences between all pairs of means are set out in Table 28. Here, for example, the difference between the mean of nitrogen deficient plants and the mean of magnesium deficient plants is 84 mg. This is less than 204 mg, so there is no significant difference between these treatments in their effects on dry weight. The only differences which exceed 204 mg are those in the top row. Plants grown in the complete solution are significantly heavier than those grown in mineral deficient solutions.

The table shows that these plants are significantly heavier at the $0\cdot1\%$ level. The minimum necessary difference can be calculated as before, but taking t from the column headed $p = 0.001$. For ten degrees of freedom, $t = 4.59$, and the least difference is $4.59 \times 91.37 = 419$ mg. All differences in the top row are greater than this.

Now re-examine the original data of Table 25. It is clear that the complete solution does produce the heaviest plants, but it also appears that the plants lacking micro-nutrients were heavier than those lacking major mineral elements, and that those lacking magnesium are distinctly the lightest of all. How is it that these clear differences did not show as significant in the analysis?

One important assumption made in an analysis of variance is that the variance is *uniform* throughout the whole body of data being analysed. It is obvious from Table 25 that this is not true of the data in this experiment. The complete solution produced heavy plants, ranging from 750 mg to 1172 mg in weight; the smaller plants in mineral deficient solutions *could not possibly* have such a wide range of weight, since every one of them weighed less than 300 mg. The plants grown in complete solution have a relatively large variance, and when analysing the variance the effects of this large variance have been spread out over *all* treatments. This has set too high a value for the least significant difference between means of the mineral deficient treatments. Their variance is actually much less, and one can be satisfied with relatively small differences between their means. The values of the standard deviation of each treatment, given in Table 25 confirm what has just been said. Though these are only rough estimates of standard deviation, being based on only three observations each, they do show how great is the standard deviation (and also the variance) of the complete treatment, when compared with that of the others. This extra large standard deviation is swamping the analysis of the finer differences between the other treatments.

The last column of figures in Table 25 shows that, within reasonable limits, the figure obtained by dividing the standard deviation of each treatment by the mean of that treatment is always very much the same. In other words, for each treatment there is an almost constant ratio between its standard deviation and its mean, the standard deviation being about one-fifth of the mean. In a situation like this, one cannot rely on the results of an analysis of variance; perhaps this accounts for the unsatisfactory results of the earlier calculations.

The way round this difficulty is to take the logarithm of each observation. This is termed making a **transformation**. The logarithmic transformation is used when the standard deviation is proportional to the mean for each treatment. The effect of this transformation is to distort the set of observations in the manner shown in Fig. 15, but this distortion has been effected according to a regular system. The scale shown in the bottom diagram is like that found on a slide-rule. This logarithmic scale, when compared with the usual (arithmetic) scale appears to have been stretched at its lower (left) end,

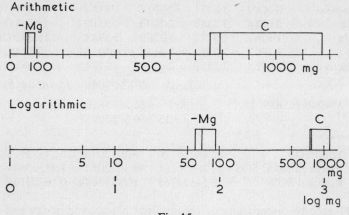

Fig. 15

and compressed at its upper (right) end. The positions of the three sets of results marked on the arithmetic scale have been carried along during this stretching and compressing, still keeping their same order, but now with different distances between them. The three 'complete' values were widely spaced (large variance) on the arithmetic scale; on the logarithmic scale they lie closer together, due to compression of this part of the scale. At the other extreme, the three ' – Mg' values, formerly very close together, have now been spaced further apart. On the arithmetic scale the ranges of the sets of figures were very unequal; on the logarithmic scale these ranges are more-or-less equal to one another. The range appears to be the same for all three sets of figures, so it is reasonable to assume that the variance too is the same for all treatments. By making this transformation one has eliminated the inequality between variances, and may now perform an analysis of variance, with reliable and useful results.

In Table 29 the data of this example are shown after logarithmic transformation. The log dry weights are found by looking them up in an ordinary table of logarithms. The analysis then proceeds as before, treating the logged data just as if they were ordinary data. Four-figure logarithms produce eight decimal places when squared, so this is an instance where a calculating machine is essential. It is

107

Table 29

Calculation of sums of squares of data of Table 25, after logarithmic transformation

Culture solution	Log dry weight of replicates				Sums of squares
	1	2	3	Totals	
Complete	3·0689	2·8751	2·8943	8·8383	26·0613 1971
– Ca	2·0334	2·2380	2·2405	6·5119	14·1631 9981
– Mg	1·8261	1·9777	1·7709	5·5747	10·3820 2531
– N	2·1703	2·3692	1·9638	6·5033	14·1798 21 17
– micro-nut	2·4728	2·3856	2·4200	7·2784	17·6622 2720

$$\text{Totals} \quad G = 34\cdot7066 \qquad H = 82\cdot44859320$$

$$K = 246\cdot86568604 \qquad K/3 = 82\cdot28856201$$
$$G^2/15 = 80\cdot30320557$$

$$u = 3 \qquad v = 5$$

Total SOS $= 82\cdot44859320 - 80\cdot30320557 = 2\cdot14538763$
'Between rows' SOS $= 82\cdot28856201 - 80\cdot30320557 = 1\cdot98535644$
Residual SOS $= 2\cdot14538763 - 1\cdot98535644 = 0\cdot16003119$

NOT safe to discard any of the decimal places, for in the end the residual sum of squares is a small difference between relatively large numbers. Rounding off to fewer decimal places might introduce serious error into the estimate of the residual mean square.

Table 30

Analysis of variance of the data from Table 29

Source of variance	Sum of squares	Degrees of freedom	Mean square
Treatments (between rows)	1·98535644	4	0·49633911
Residual	0·160 03119	10	0·016 003119
Totals	2·14538763	14	

The analysis of variance is shown in Table 30, the numbers of degrees of freedom being the same as before. The variance ratio for the effect of treatment is:

$$F = \frac{0\cdot4963}{0\cdot01600} = 31\cdot01$$

For $n_1 = 4$, $n_2 = 10$, and $p = 0\cdot001$, the table gives $F = 11\cdot3$ (as before, of course). The effect of treatment is therefore significant at the 0·1 % level.

The standard deviation of the difference between any pair of means is based on the estimate on the residual mean square. As mentioned before, this method is not strictly valid, and may lead one to count as significant some differences which are not significant, but it is a useful guide.

$$\sigma_d = \sqrt{\left(2 \times \frac{0 \cdot 016}{3}\right)} = \sqrt{(0 \cdot 0107)} = 0 \cdot 1034$$

From the table, for ten degrees of freedom:

For $p = 0 \cdot 05$, $t = 2 \cdot 23$, so the difference of means

must exceed $2 \cdot 23 \times 0 \cdot 1034 = 0 \cdot 2306$

For $p = 0 \cdot 01$, $t = 3 \cdot 17$, so the difference of means

must exceed $3 \cdot 17 \times 0 \cdot 1034 = 0 \cdot 3278$

For $p = 0 \cdot 001$, $t = 4 \cdot 59$, so the difference of means

must exceed $4 \cdot 59 \times 0 \cdot 1034 = 0 \cdot 4746$

The means of the treatments are given in Table 31. One point of interest is that the mean for ' $- Ca$ ' is less than the mean for ' $- N$ ', whereas in the original data the mean for ' $- Ca$ ' was greater than the mean for ' $- N$ '. This is because one has added the logarithms and divided by 3, which is equivalent to multiplying the original three dry weights together and taking their cube root. These logarithmic means are in fact the logarithms of the *geometric* means of the original data. When dealing with data for which a logarithmic transformation is appropriate, the geometric means are a more valid estimate of treatment effects than are the usual arithmetic means, which are more familiar. The geometric means for each treatment, obtained by taking the antilogarithm of each of the mean logarithms,

Table 31

Means of logarithms of dry weights (data from Table 29)

Culture solution	Mean log dry weight	Geometric mean dry weight (mg)
Complete	2·9461	883
– Ca	2·1706	148
– Mg	1·8582	72
– N	2·1678	147
– micro-nutrients	2·4261	267

are given in the third column of Table 31. Having now taken antilogarithms the calculations are back to the original scale of measurement, and the geometric means are in the original units, milligrams.

To consider differences between means one must return to the mean logarithms. The differences between these are shown in Table 32. Comparing Table 32 with Table 28, one now finds that five

Table 32

Differences of logarithmic means, and their significance

Greater mean	*Lesser mean*			
	– Mg	– N	– Ca	– Micro-nutrients
Complete	1·0879***	0·7783***	0·7755***	0·5200***
– micro-nutrients	0·5679***	0·2583*	0·2555*	—
– Ca	0·3124*	0·0028	—	—
N	0·3096*	—	—	—

Levels of significance: *** $p = 0.001$

* $p = 0.05$

The remaining difference is not significant.

more differences are significant. The only difference which is not significant is that between ' – Ca' and ' – N', which was the one which had become reversed as a result of transformation. By using this transformation one has been able to achieve a greater resolution between the means of lower magnitude, and has demonstrated the significance of differences which did not show in the original analysis.

Finally a word or two of caution: the logarithmic transformation must only be used when the standard deviation is proportional to the mean. To use it at other times (especially as a desperate attempt to get 'significant' results, when the ordinary analysis has failed to give them) will lead to false conclusions. Also, the logarithmic transformation takes longer than the ordinary analysis, and there are far more opportunities for making mistakes in calculation, so it should be avoided if possible. When results are clear-cut, they will show in the ordinary analysis and there is little to be gained by transformation. Occasionally, as in this special example, the transformation is beneficial and worth undertaking.

4 *A use for the 2 × 2 contingency table in ecology*

In a given community or habitat, two species may show association, that is to say, the presence of one species makes it more likely that the other species will be present too. This may happen because both species are similarly affected by microclimatic influences (one shade-plant will tend to be associated with another shade-plant), or by soil conditions (calcicolous plants will tend to occur in association), or it may be that one species provides favourable conditions for the others (parasite and host, insect larvae on certain types of food plant) or for many other reasons. In many instances, the association is virtually complete and needs no statistical backing. For example the shade-plant, *Geum urbanum*, is associated with the trees and bushes which provide it with shade. It is abundant beneath them, but is seldom found outside the shaded zone. In other instances, association may not be so marked, and to demonstrate it, one must sample the community and analyse the result statistically. The usual procedure is to lay down quadrats of suitable size (see p. 114) within the area under investigation. The presence or absence of each species is then recorded for every quadrat.

Before studying an actual example, consider the imaginary example illustrated by Table 33. Species A occurs within the area to the extent

Table 33

Contingency tables in an imaginary example (see text)

a. No association

		SPECIES A		
		Present	*Absent*	*Totals*
SPECIES B	*Present*	25	75	100
	Absent	25	75	100
	Totals	50	150	200

b. Positive association

		SPECIES A		
		Present	*Absent*	*Totals*
SPECIES B	*Present*	45	55	100
	Absent	5	95	100
	Totals	50	150	200

c. Negative association

		SPECIES A		
		Present	*Absent*	*Totals*
SPECIES B	*Present*	10	90	100
	Absent	40	60	100
	Totals	50	150	200

that the chances of it appearing in a given quadrat are 1 in 4, so it is present in 50 quadrats out of the 200 recorded. Species B is commoner than this, being found in half of the quadrats. If the two species are not associated one would expect the results of Table 33a. Of the 50 quadrats containing A one would expect B to be present in half (25) and absent from half (25). Similarly of the 150 quadrats without A, one would expect B to be present in half (75) and absent from half (75).

Suppose that there is association between A and B. The presence of A in a quadrat makes it more probable that B will occur there too. Conversely the presence of B increases the probability of the presence of A. This is shown in Table 33b, where there is a disproportionately greater number of quadrats containing both species or neither species. The quadrats in which A or B occur alone are fewer in number. The degree of association determines the extent to which the figures of Table 33b differ from those of Table 33a, and the statistical significance of this difference can be tested by the χ^2 test.

Another possibility is *negative* association. The presence of one species may make it *less* likely that the other will be present also. If there is negative association the table might look like Table 33c. There are relatively more quadrats containing one or other of the species alone, and fewer quadrats containing both or neither. Again the χ^2 test can be used to establish significance.

Table 34

Contingency table, showing occurrence of two plant species in quadrats in a blanket-bog community

SPECIES A : *Calluna vulgaris*

		Present	Absent	Totals
SPECIES B:	*Present*	90	181	271
Eriophorum		(94)	(177)	
vaginatum				
	Absent	66	113	179
		(62)	(117)	
	Totals	156	294	450

Terms in parentheses are values expected on the null hypothesis that there is no association between the occurrence of the two species.

The data given in Table 34 refer to a survey made in a blanket-bog community, using 450 randomly placed quadrats, each 1 metre square, and recording the presence or absence of two species, *Calluna vulgaris* and *Eriophorum vaginatum*. The numbers expected on the null hypothesis of there being no association are shown in brackets. The figures for 'both A and B present' and for 'neither A nor B present' are lower than would be expected on the null hypothesis. This is like Table 33c, representing negative association. The discrepancy between observation and expectation is only four quadrats in each cell of the table. How often would one expect a difference of this amount to arise by chance? χ^2 is found by the short-cut equation given on p. 177:

$$\chi^2 = \frac{450\,(90 \times 113 - 181 \times 66 - 450/2)^2}{271 \times 179 \times 156 \times 294} = 0\cdot810$$

In a 2×2 contingency table there is one degree of freedom, and the table of χ^2, for $p = 0\cdot5$, gives $\chi^2 = 0\cdot455$, and for $p = 0\cdot2$, gives $\chi^2 = 1\cdot64$. Thus the probability of the observed value of χ^2 lies between $0\cdot2$ and $0\cdot5$. This means that in 20% to 50% of surveys one would expect to obtain similar results in the *absence* of any association (positive *or* negative). So there is no evidence to lead to rejection of the null hypothesis; no association has been demonstrated.

Though the results have been inconclusive one could consider repeating the survey to obtain fresh and decisive data. It can be assumed that recording has been carefully done, and since there is no difficulty in deciding between presence or absence, there is probably no scope for improving the accuracy of recording each quadrat. One might decide to repeat the survey, using a greater number of quadrats, say 1000 or 1500. This *might* give a significant indication of association, but before embarking on such an extensive scheme it is worth considering another relevant factor—the size of quadrat used for sampling.

Suppose that the quadrats had been 10 cm square. With plants of the size of *Calluna* and *Eriophorum* it would most frequently happen that if a quadrat contained *one* plant of either of these species there would be *no room* left for a plant of the other species. Apparently there would be strong negative association, but this would have arisen solely because the small size of the quadrat made it very improbable that a quadrat would contain more than one plant. This apparent association is merely a consequence of the sampling technique and does not represent an association, in the ecological sense. Had one been dealing with an association between moss and lichen species growing on a wall, or limpets and microscopic algae on a seashore rock face, a 10 cm square quadrat would probably be very satisfactory, for in these cases the size of the organisms and the general scale of the pattern of their distribution is smaller than the dimensions of the quadrat.

At the other extreme, suppose that the quadrats used had been 10 m square, or even larger. With these quadrats it would be seldom that one would fail to enclose at least one plant of each species. Nearly every quadrat would show the presence of both species and the table would apparently indicate strong positive association. Again, this artefact is the product of the sampling technique. The quadrat is larger than the scale of the pattern of the distribution of the two species. Thus one can obtain results ranging between a strong negative association and a strong positive association, depending on the size of quadrat used. Over a range of intermediate sizes one may fail to demonstrate any association whatever.

From this, it follows that investigation of association should make use of quadrats of several different sizes. By examining the results at each size, it will be possible to discount the effects of quadrats too large or too small for the particular species being surveyed. In the example, the quadrats were 1 m square. In the absence of other data one cannot tell what would be the effect of increasing or decreasing the size. Having regard to the average size of plants of these two species, it is probable that quadrats 2 m square would be worth trying. This way of continuing the survey would probably be more fruitful than amassing data for a further 500 or 1000 quadrats of the original size, for if size is influencing the results, it will continue to do so, no matter how many quadrats are counted, and the results from 1500 may be no more significant than these.

In this example, the quadrats were randomly placed within the area to be sampled. This has some advantage if one wishes to obtain a quantitative measure of the degree of association, using methods not described in this book. If one wishes simply to know whether or not there is association, random quadrats are not necessary, in fact some systematic arrangement is preferable. One of several possible ways of arranging the quadrats is shown in Fig. 16, using a belt of quadrats or a number of such belts, extending across the area to be sampled. Records are taken for each small quadrat, which may be, say, 50 cm square. These results can be analysed for association.

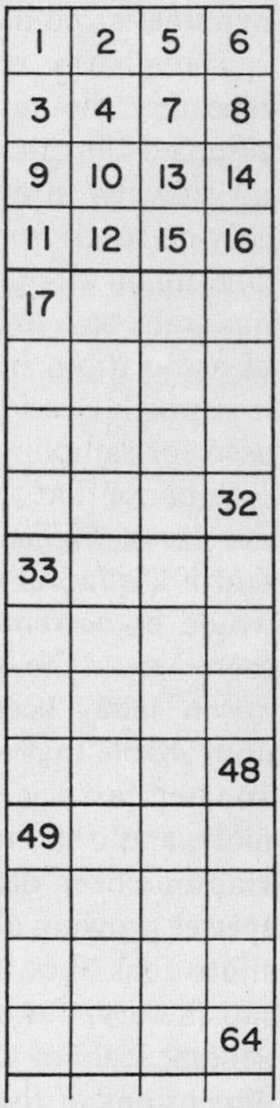

1	2	5	6
3	4	7	8
9	10	13	14
11	12	15	16
17			
			32
33			
			48
49			
			64

Fig. 16

Then one obtains the results for 1 m square quadrats, by taking the smaller ones in groups of 4 (1 to 4, 5 to 8, 9 to 12, and so on) and combining their results. One can also obtain data for quadrats 2 m square by combining the small quadrats in groups of 16 (1 to 16, 17 to 32, 33 to 48, and so on). A planned systematic arrangement makes it possible to obtain data for quadrats of various sizes from a single survey. Though shown in Fig. 16 as squares, quadrats need not be square; they can be rectangular, and be amalgamated into larger rectangular or square groups.

The discussion of this example has raised two points of general relevance:

1. Because of the complexity of the patterns of the distribution of living organisms, there are problems in ecological sampling which do not arise in other branches of biology. Thus it is essential to describe sampling technique in detail when writing up results. A common-sense approach will generally produce a workable and reliable sampling technique, but for extensive and critical investigations it is worth while to consult a text which deals with ecological sampling in detail. The texts by Greig-Smith and by Kershaw, listed on p. 205, are recommended.

2. In the midst of calculations and equations, one must always keep in mind the relationships between the figures on paper, and the living material which the figures represent.

5 *Another genetical example*

A cross was made between two pure-breeding strains of radish, *Cherry Belle* (round, red radish), and *White Icicle* (long, white radish). Seeds were collected, and sown the following year to give the F_1 generation. All these plants had a radish which was purple in colour, and in shape was intermediate between long and round. These plants were allowed to flower and interpollinate, and seeds were collected. In the following year, 160 seeds were sown, to discover in what ratio and in what combinations the various shapes and colours would occur. The observed results are shown in Table 35. The expected results in this table are based upon the scheme of inheritance illustrated in Table 36, which is consistent with the results obtained in the F_1 generation, and establishes a hypothetical ratio for the F_2 generation. This assumes that colour and shape are governed by a pair of alleles each, with intermediate expression in the heterozygote, and that there is no linkage between genes for colour and those for shape.

Since it was intended to test the observed ratio by the χ^2 test, it was necessary to ensure that there would be at least five plants in

Table 35

χ^2 **test on results of radish-breeding experiment**

Shape	Round			Intermediate			Long			
Colour	Red	Purple	White	Red	Purple	White	Red	Purple	White	Totals
Observed										
No. (O)	14	11	12	17	36	23	11	21	15	160
Expected										
No. (E)	10	20	10	20	40	20	10	20	10	160
$O - E$	4	−9	2	−3	−4	3	1	1	5	0
$(O - E)^2$	16	81	4	9	16	9	1	1	25	
$(O - E)^2/E$	1·60	4·05	0·40	0·45	0·40	0·45	0·10	0·05	2·50	

$$\chi^2 = 10\cdot00$$

Table 36

Theoretical basis of radish-breeding experiment

PARENTAL TYPES	Cherry Belle (round, red) $s^r s^r c^r c^r$	×	White Icicle (long, white) $s^l s^l c^w c^w$

Gametes $s^r c^r$ ⟍ ⟋ $s^l c^w$

F$_1$ GENERATION   intermediate, purple

$s^r s^l c^r c^w$

Gametes

	$s^r c^r$	$s^r c^w$	$s^l c^r$	$s^l c^w$
$s^r c^r$	$s^r s^r c^r c^r$ round, red	$s^r s^r c^r c^w$ round, purple	$s^r s^l c^r c^r$ inter., red	$s^r s^l c^r c^w$ inter., purple
$s^r c^w$	$s^r s^r c^w c^r$ round, purple	$s^r s^r c^w c^w$ round, white	$s^r s^l c^w c^r$ inter., purple	$s^r s^l c^w c^w$ inter., white
$s^l c^r$	$s^l s^r c^r c^r$ inter., red	$s^l s^r c^r c^w$ inter., purple	$s^l s^l c^r c^r$ long, red	$s^l s^l c^r c^w$ long, purple
$s^l c^w$	$s^l s^r c^w c^r$ inter., purple	$s^l s^r c^w c^w$ inter., white	$s^l s^l c^w c^r$ long, purple	$s^l s^l c^w c^w$ long, white

F$_2$ GENERATION   1 round red: 2 round purple: 1 round white: 2 intermediate red: 4 intermediate purple: 2 intermediate white; 1 long red: 2 long purple: 1 long white.

s^r, s^l = alleles for shape (round, and long)

c^r, c^w = alleles for colour (red, and white)

every category. This condition was in mind when the seeds were sown, and it was reckoned that with 160 plants there would be approximately ten in each of the smallest categories, so amply satisfying the minimum condition for the χ^2 test.

The calculation of χ^2 is shown in Table 35. With only random variation, the value of χ^2 would be that given in Table 80, p. 198. There for eight degrees of freedom (nine categories; one restriction, that the total must be 160) it is shown that for $p = 0.20$, $\chi^2 = 11.03$, and for $p = 0.50$, $\chi^2 = 7.34$. The probability of obtaining a value of χ^2 of 10.00 thus lies between 0.20 and 0.50. The observed ratio does not deviate significantly from the expected ratio, and the observed results are therefore consistent with the scheme of inheritance outlined in Table 36.

6 *Dealing with proportions and percentages*

In a beechwood, 152 snails of the species *Cepea nemoralis* were collected; 45 of these had banded markings on the shell. Under several hedges, a sample of 64 snails of the same species was collected; 57 of these had banded markings. These two samples can be compared only if the number of banded shells is expressed as a proportion or as a percentage of the whole. In the wood, the proportion of banded shells was $45/152 = 0.296$. As a percentage, this is 29.6%. In the hedgerows, the proportion was $57/64 = 0.891$, or 89.1%. Comparison of these proportions or percentages shows the selective advantage of the banding in the hedgerows, where the background is of variable colours.

The significance of this result could be tested by a 2×2 contingency table, using the χ^2 test to compare the numbers of banded and unbanded snails found in the two habitats. This gives $\chi^2 = 66.3$, which well exceeds the tabulated value of χ^2, when $p = 0.001$, with 1 degree of freedom. There is no doubt that there is a significant difference between the proportions of banded snails in the two habitats. This knowledge may be all that is required, and the analysis will be complete. However, it might be necessary to obtain a numerical estimate of the proportion of banded snails in the two habitats. To do this, several samples must be collected from each habitat, the proportion of banded snails calculated for each sample, and the mean of these proportions calculated for each habitat. For instance, four samples from the wood might contain banded snails in the proportions 0.296, 0.315, 0.265, and 0.288. These figures are values of a variable, the proportion of banded snails, and they can be analysed in several ways, according to what information is required. The only difficulty is that all the methods of analysis discussed in this book depend upon the variable having a normal distribution. The distribution of a set of proportions or of percentages is usually not normal. This may make little difference if the results are clear-cut; analysis by the usual methods will give reliable conclusions. When the results are not so distinct, the exact form of their distribution becomes of greater importance.

When the proportions are scattered around 0·5 (or percentages around 50%) their distribution is often approximately normal, but difficulties arise at both ends of the scale. Proportions cannot be less than 0 or greater than 1, and similarly all percentages must lie in the range 0% to 100%. By contrast, it is assumed in the normal distribution that the tails of the curve extend to infinity on either side of the mean. The consequence of this is that, if a sample has a mean of say, 0·9, it will tend to have a smaller standard deviation than a sample with a mean of 0·5, simply because the observations greater than 0·9 cannot be greater than 1·0. They are confined by the end of the scale, and cannot deviate greatly from 0·9. The nearer the mean lies to 1·0 the more pronounced this effect becomes. The same thing happens at the other end of the scale. Thus the data will not conform to the normal distribution. Instead of having a fixed value for the standard deviation, there will be a maximum value of standard deviation for proportions around 0·5, and decreasing values of standard deviation at either end of the scale. This phenomenon of a regularly varying standard deviation was first met in Example 3, p. 103, and in that example the solution was to make a logarithmic transformation. In this example the standard deviation varies in a different manner, and a different transformation is required, the **arcsin transformation** (Fig. 17).

To transform a proportion, of value x, one finds an angle θ for which $\sin \theta = \sqrt{x}$. For example, if $x = 0·64$, one first takes its square root, 0·8, and then looks in a table of natural sines to find an angle between 0° and 90° of which the sine is 0·8. Sin 53° 8′ = 0·8, and so, working in decimal fractions rather than in minutes of angle, the required angle is 53·13°. This is the transformed value (θ) used in an analysis, instead of x. Working with percentages, one follows the same procedure, but divides by 100 to convert the percentage to a proportion before beginning the transformation. For example, 40% corresponds with 0·4, $\sqrt{0·4} = 0·63246$, sin 39·23° = 0·6325, so $\theta = 39·2°$.

To save time in transforming, special tables have been prepared. In some texts, values are tabulated to several decimal places, but for most work such accuracy is not needed. The arcsin transformation is given in Table 81, p. 199. The table can be used for proportions by entering at 100 times the value of the proportion. It does not matter in what unit the angle θ is expressed. In Table 81, angles are in degrees, ranging from 0 to 90. In some other books of tables, the angles are in radians, so the values range between 0 and 1·571 radians ($=90°$). One must keep to one unit or the other during the course of an analysis.

Another way in which percentages may arise in biological work is in ecology, for the estimation of frequency, cover index, valence, and the like. Some cover values, expressed as percentages are given in Table 37. A transect was laid across marshy ground and, at pre-determined points along the transect, the percentage cover values of

Table 37

Cover values, as percentages, and with arcsin transformation

Sample No.	Percentage cover value		Transformed (degrees)	
	Agrostis	*Juncus*	*Agrostis*	*Juncus*
1	60	15	50·8	22·8
2	52	12	46·2	20·3
3	63	26	52·5	30·7
4	51	22	45·6	28·0
5	43	23	41·0	28·7
6	52	35	46·2	36·3
7	38	37	38·1	37·5
8	56	52	48·5	46·2
9	44	49	41·6	44·4
10	23	50	28·7	45·0
11	40	60	39·2	50·8
12	23	70	28·7	56·8
13	32	73	34·5	58·7
14	31	92	33·8	73·6
15	14	94	22·0	75·8
16	20	100	26·6	90·0

two plants species were estimated, for 5-cm lengths of the transect. Sixteen such lengths were studied, and the cover values for each are tabulated. The fourth and fifth columns of this table contain the transformed values of these percentages. The effect of this transformation is shown in Fig. 17, where the 16 observations for *Juncus articulatus* are marked on the scales.

To find the mean cover value for *Juncus*, sum the *transformed* values and divide by 16:

$$(22·8 + 20·3 + \ldots + 75·8 + 90·0) / 16 = 745·6/16 = 46·6 \text{ degrees}$$

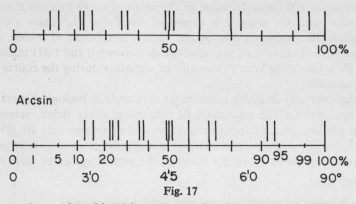

Fig. 17

From the arcsin table, this corresponds with a mean of 53%.

Compare this with the mean of the raw, untransformed figures:

$$(15 + 12 + \ldots + 94 + 100) / 16 = 810/16 = 50 \cdot 6\%$$

The means of the raw data and the transformed data are not equal, as was found for the data of the mineral nutrient experiment (p. 109) after logarithmic transformation. The appropriate mean for this analysis is that derived from the arcsin transformation; for *Juncus*, the mean cover value is 53%.

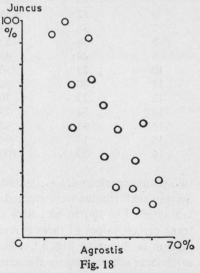

Fig. 18

The observations recorded in Table 37 were collected with the aim of discovering a correlation between the cover value of *Agrostis stolonifera* and that of *Juncus articulatus*. In such a series as this, where correlation is obvious from the scatter diagram of the untransformed observations (Fig. 18) it is hardly worth while to bother with the transformation, but for the sake of exemplifying the technique, the transformed data will be used. The stages in calculation are shown in Table 38. Values of x^2 and y^2 can be found in the table of squares (Table 72, p. 189), but xy must be obtained by long multiplication.

120

Table 38

Calculation of correlation coefficient and regression for transformed cover values

Sample No.	Cover values (degrees)		x^2	y^2	xy
	Agrostis x	Juncus y			
1	50·8	22·8	2 580·64	519·84	1 158·24
2	46·2	20·3	2 134·44	412·09	937·86
3	52·5	30·7	2 756·25	942·49	1 611·75
4	45·6	28·0	2 079·36	784·00	1 276·80
5	41·0	28·7	1 681·00	823·69	1 176·70
6	46·2	36·3	2 134·44	1 317·69	1 677·06
7	38·1	37·5	1 451·61	1 406·25	1 428·75
8	48·5	46·2	2 352·25	2 134·44	2 240·70
9	41·6	44·4	1 730·56	1 971·36	1 847·04
10	28·7	45·0	823·69	2 025·00	1 291·50
11	39·2	50·8	1 536·64	2 580·64	1 991·36
12	28·7	56·8	823·69	3 226·24	1 630·16
13	34·5	58·7	1 190·25	3 445·69	2 025·15
14	33·8	73·6	1 142·44	5 416·96	2 487·68
15	22·0	75·8	484·00	5 745·64	1 667·60
16	26·6	90·0	707·56	8 100·00	2 394·00
Totals	632·9	745·6	25 601·21	40 852·02	26 838·60
	Σx	Σy	Σx^2	Σy^2	Σxy

From Table 38, the following quantities may be calculated:

$\Sigma d_x^2 = 25\,601\cdot21 - (623\cdot9)^2/16 =$

$$25\,601\cdot21 - 24\,328\cdot19 = \quad 1\,273\cdot02$$

$\Sigma d_y^2 = 40\,852\cdot02 - (745\cdot6)^2/16 =$

$$40\,852\cdot02 - 34\,744\cdot94 = \quad 6\,107\cdot08$$

$\Sigma d_x d_y = 26\,838\cdot60 - (623\cdot9 \times 745\cdot6)/16 =$

$$26\,838\cdot60 - 29\,073\cdot69 = -2\,235\cdot09$$

The negative sign of $\Sigma d_x d_y$ indicates a negative correlation, and its significance is estimated by calculating r:

$$r = \frac{-2\,235\cdot09}{\sqrt{(1\,273\cdot02 \times 6\,107\cdot08)}} = \frac{-2\,235\cdot09}{\sqrt{(7\,774\,434\cdot9816)}} = \frac{-2\,235\cdot09}{2\,788\cdot2} = -0\cdot802$$

For 15 degrees of freedom, and $p = 0\cdot001$, Table 79 gives $r = 0\cdot725$. The correlation is significant at the $0\cdot1\%$ level.

The next step is to calculate the regression equations:

E

Statistics for biology

For the regression of y on x:

$$\bar{x} = 623 \cdot 9/16 = 38 \cdot 99 \qquad\qquad \bar{v} = 745 \cdot 6/16 = 46 \cdot 60$$

$$b = \frac{-2235 \cdot 09}{1273 \cdot 02} = -1 \cdot 756$$

$$\therefore \; y = 46 \cdot 60 + (-1 \cdot 756)(x - 38 \cdot 99) = 46 \cdot 60 - 1 \cdot 756x + 68 \cdot 47$$
$$\therefore \; y = 115 \cdot 07 - 1 \cdot 756x$$

For the regression of x on y:

$$b = \frac{-2235 \cdot 09}{6107 \cdot 08} = -0 \cdot 366$$

$$\therefore \; x = 38 \cdot 99 + (-0 \cdot 366)(y - 46 \cdot 60) = 38 \cdot 99 - 0 \cdot 366y + 17 \cdot 06$$
$$\therefore \; x = 56 \cdot 05 - 0 \cdot 366y$$

The two regression lines are drawn in Fig. 19. They both slope steeply downwards to the right, showing a negative correlation. The interpretation of this is that the percentage cover of the two species is

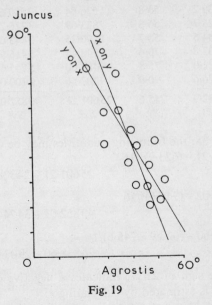

Fig. 19

inversely correlated; when the percentage cover of *Agrostis* is high, that of *Juncus* is low, and the converse. Without further information it is not possible to comment on the ecological significance of this, for instance whether this is the effect of mutual exclusion from the small length of transect sampled, or whether it is due to varying ecological factor or factors which act differentially upon the two species.

122

7 Rate of Growth

One hundred thalli of the water-plant, *Lemna minor*, the duckweed, were placed in each of two large shallow vessels containing mineral nutrient solution. The plants were illuminated by artificial light, those in one vessel receiving double the light intensity received by those in the other vessel. Every day the number of thalli in each vessel was counted. The results are given in Table 39, and illustrated by Fig. 20.

Table 39

Vegetative multiplication of *Lemna*

Time (days)	No. of thalli	Log No. of thalli				No. of thalli (double light)
x	N	y	x^2	y^2	xy	
0	100	2·0000	0	4·00000000	0	100
1	123	2·0899	1	4·36768201	2·0899	125
2	152	2·1818	4	4·76025124	4·3636	169
3	182	2·2601	9	5·10805201	6·7803	246
4	222	2·3424	16	5·48683776	9·3696	336
5	294	2·4683	25	6·09250489	12·3415	436
6	369	2·5670	36	6·58948900	15·4020	586
7	460	2·6628	49	7·09050384	18·6396	867
8	590	2·7709	64	7·67788681	22·1672	1090
9	669	2·8254	81	7·98288516	25·4286	—
Totals 45		24·1686	285	59·15609272	116·5823	

Lemna reproduces vegetatively by budding off small thalli, which soon become detached and independent, and in their turn bud off more thalli. This manner of reproduction is closely linked with growth, and is analogous to the binary fission of bacteria, and the budding of yeasts. The progeny immediately commence to grow and bud, producing new individuals at an ever-increasing rate, until shortage of food or the accumulation of metabolic waste products has a deleterious effect. In the early phase of growth, when there are no such limitations, the number of individual thalli increases by compound interest. This experiment is designed to estimate the rate of compound interest, and to find out how it varies with light intensity. By looking at the scatter diagram of Fig. 20, it can be seen that the rate of interest is higher at the higher light intensity, but one cannot give any precise values.

123

It can be shown that, until limiting factors begin to operate, growth of this sort can be expressed by an equation:

$$N_t = N_o e^{gt}$$

where N_t = number of thalli present after a given time, t

N_o = number of thalli present originally

e = the exponential function = $2 \cdot 718$

g = the rate of increment (rate of interest) expressed as a fraction (for example, $g = 0 \cdot 2$ means that in a given time the number of thalli increases by 20%).

The growth or multiplication of many kinds of organisms follows this equation, and the methods used for tackling this example can therefore be applied also to the measurement of growth rates of bacteria, yeasts, and many higher organisms. The quantity of particular interest is g, the rate of increment.

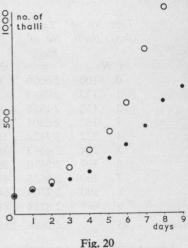
Fig. 20

Table 39 presents two variables for each observation—the number of thalli, and the time at which the observation was made. This is the first example in which time has been a variable, but there is no reason why it should not. The first stage in the analysis is to examine the data for a correlation between time and the number of thalli, and the next stage is to calculate the regression equation for number of thalli (N) on time (x). In previous examples of correlation, both variables have been subject to random error, and have been normally distributed, or approximately so. In this example, only one of the variables (N) is subject to random error. In effect, time is measured exactly. If the thalli are counted at the same hour each day, inaccuracies of the order of five or ten minutes may be ignored. This fact does not make any difference to the method of calculation, but it means that there can be only one regression equation, the regression of N on x; the regression of x on N is meaningless.

In previous examples, two straight lines drawn through the cloud of points on the scatter diagram have represented the best lines for the regression equations. It is clear from Fig. 20 that the points in this example do not lie on a straight line, either at high or at low light intensity. They lie almost perfectly on two curved lines. An attempt to calculate a best straight line for the regression of N on x is doomed from the outset. However, theory (p. 124) suggests that these lines probably have the equation, $N_t = N_o e^{gt}$, each line having its own value for g. Equations like these cannot be derived directly from the usual technique for regression equations, and the problem must be tackled from a different aspect.

Taking logarithms of both sides of the growth equation, the equality will be preserved, so:

$$N_t = N_o e^{gt}$$

gives:

$$\log N_t = \log N_o + gt \, . \, \log e$$

This can be recognised as the equation of a straight line, with a gradient, $g \, . \, \log e$. If the gradient of the line can be discovered, one can then calculate g, for $\log e$ is a constant. If this line could be expressed as a regression equation, the gradient would be the quantity, b.

The argument above shows that one will do best to take logarithms of the numbers of thalli; and to use this transformed data in calculating the regression equation, as has been done in Table 39.

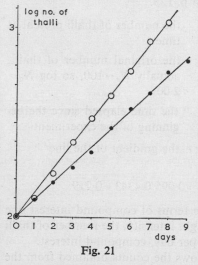

Fig. 21

Transformations were made in Examples 3 and 6, but for a reason different from that operating in this example. In Examples 3 and 6, the data were not normally distributed; transformation normalised the data so that the usual statistical techniques could be applied. In this example, the data have normally distributed error, probably, but there is a special reason to believe that they conform to a mathematical relationship for which the straight line of a usual regression equation will be useless. The data are to be transformed, according to an equation derived from the theory of growth rates, so that they will come to lie on a straight line. The equation for this line may then be estimated by the usual methods.

The transformation and calculation are given in Table 39, and the scatter diagrams of the transformed data are shown in Fig. 21. Note that the time variable has not been transformed—the equation does not require this. From the figures of Table 39:

$$\Sigma d_x = 285 - (45)^2/10 = 285 - 202{\cdot}5 = 82{\cdot}5$$

$$\Sigma d_y = 59{\cdot}1561 - (24{\cdot}1686)^2/10 = 59{\cdot}1561 - 58{\cdot}4121 = 0{\cdot}7440$$

$$\Sigma d_x d_y = 116{\cdot}5823 - (45 \times 24{\cdot}1686)/10 = 116{\cdot}5823 - 108{\cdot}7587 = 7{\cdot}8236$$

It is hardly necessary to calculate r, but for the sake of completeness:

$$r = \frac{7{\cdot}8236}{\sqrt{(82{\cdot}5 \times 0{\cdot}7440)}} = \frac{7{\cdot}8236}{\sqrt{(61{\cdot}38)}} = 0{\cdot}999$$

Highly significant at the 0·1 % level; in fact, almost certainly true. The regression of y on x is found as follows:

$$\bar{x} = 45/10 = 4{\cdot}5 \qquad\qquad \bar{y} = 24{\cdot}1686/10 = 2{\cdot}41686$$

and the gradient, $b = \dfrac{7{\cdot}8236}{82{\cdot}5} = 0{\cdot}095$

$$\therefore\ y = 2{\cdot}41686 + 0{\cdot}095(x - 4{\cdot}5)$$
$$\therefore\ y = 1{\cdot}99 + 0{\cdot}095x$$

Compare this with the equation on p. 125.

y corresponds with $\log N_t$ — the number of thalli present at time t

1·99 corresponds with $\log N_o$ — the original number of thalli; actually $N_o = 100$, so $\log N_o = 2{\cdot}00$

x corresponds with t — the time elapsed since the beginning of the experiment

and 0·095 corresponds with $g \cdot \log e$ — the gradient of the line

So $g \cdot \log e = 0{\cdot}095$

and $g = 0{\cdot}095 / \log e = 0{\cdot}095/0{\cdot}4343 = 0{\cdot}219$

This is the rate of increment. In terms of compound interest one would say that, under the single light intensity, the number of thalli is increasing at the rate of 21·9 % per day, compound interest.

The last column in Table 39 shows the counts obtained from the culture exposed to double light intensity. Note that towards the end of the experiment the number of *new* thalli produced is almost exactly double that of the other culture. The actual increment, in number of new thalli produced, is approximately double, and is thus proportional to the intensity of illumination. When this series of figures is analysed, it is found that $g = 0.306$. The *rate* of increment is only about $1\frac{1}{2}$ times the value it held under single illumination. Doubling the light intensity doubles the number of thalli produced, but surprising as it may seem on first consideration, it does not double the rate of growth.

This example has dealt with a relationship which commonly occurs in biology, and the method can be applied in several fields of study. The reader will often come across other instances in which a preliminary consideration of the underlying theory will guide the way to a useful transformation.

PROBLEMS

25 The vegetation in two areas of a wood on a steep hillside was sampled by throwing a metal ring of area 0.1 m². The sampling was done separately by three persons in each area. In the first area, at 120 m above sea level, the bramble, *Rubus fruticosus agg.* was ringed 16 times out of 50 throws by the first person, 9 times out of 25 by the second, and 19 times out of 50 by the third person. In the second area studied, at 60 m above sea level, bramble was ringed 35 times out of 50, 17 times out of 25, and 42 times out of 50, by the three investigators. Convert these figures to proportions or percentages, perform the arcsin transformation, and investigate the apparently greater occurrence of bramble at the lower level (2nd area).

26 In the investigation referred to in Problem 7 (p. 19) 50 of the samples were taken at one end of the lawn, which was under the shade of trees, while the other 50 samples were taken at the open, sunny end of the lawn. The results for daisy plants were:

No. of plants ringed		0	1	2	3	4	5	6	7	Total throws
Frequency }	Shaded	11	13	8	5	8	1	1	3	50
(no. of throws) }	Sunny	36	7	4	2	1	0	0	0	50

Transform the data by the most appropriate technique (see Table 51, p. 152), and discover if there is a significant difference between the means of the numbers of daisy plants in the shaded and sunny ends of the lawn.

27 In the same investigation the records provide evidence for or against association of the two species, daisy and ribwort. The table indicates the number of throws out of a total of 100, in which one or other, or both or neither of the two species was ringed:

		Daisy	
		Present	*Absent*
	Present	12	25
Ribwort			
	Absent	41	22

Is there any evidence of association between these species?

28 In the data of Problem *18* (p. 70), there are 29 holly leaves with an even number of spines, and 22 leaves with an odd number. Does this significantly indicate that holly leaves tend to have an even number of spines?

11 Another type of test

In an experiment to demonstrate the importance of visual feedback in the control of arm movements, a subject was asked to trace with a pencil an irregular path marked on paper. The time taken was recorded for four trials in which the subject could watch his arm and hand directly, and for four trials in which he could see only the reflection of his arm and hand in a large mirror. The results are given in Table 40.

Table 40

Times taken to trace a path (seconds)

| Direct viewing | 12 | 11 | 15 | 10 |
| Reflected viewing | 18 | 16 | 20 | 14 |

Apparently it takes longer to trace the path if the visual feedback is reversed by the mirror. One way of testing this data for statistical significance would be to calculate $\bar{x}_1$ and $\bar{x}_2$, estimate σ_1^2 and σ_1^2, and from these estimate σ_d. Then we would apply the t-test. This test will give us a reliable analysis provided that one very important assumption is true. We assume that the data are samples from populations with normal distributions. If the distributions are *not* normal, the test may lead to a false interpretation of the experiment.

With so few measurements available it is impossible to say whether the distribution of tracing-times is normal or not. Many kinds of measurements are distributed normally, so we *could* risk making the assumption for this data, but there are reasons for thinking that the distribution might be skew. There must be a sharp lower limit for the time taken to trace the path, governed by the maximum speed of drawing a line. At the other extreme, when an error has been made there will be further delay while it is being corrected, and the resulting confusion in the mind of the subject could lead to hesitation and even further errors. The distribution would have a rather long upper tail.

If the distribution is skew, the t-test will be unreliable. We cannot overcome the problem by using a transformation, because we do not know what form the distribution has, and so do not know what transformation to use. To analyse these data we need a test which does not require us to make assumptions about the distribution. Such a test is called a **distribution-free test**. In some books it is called a non-parametric test. Distribution-free tests depend on relatively simple mathematical formulae which will be explained in the next section.

Factorial numbers and combinations

Suppose we have four letters, A, B, C, and D, and four spaces in which we can write them. We can fill the first space in any one of these ways:

A ···   *or*   B ···   *or*   C ···   *or*   D ···   =4 ways

The second space can be filled in these ways:

AB ··	BA ··	CA ··	DA ··
or	*or*	*or*	*or*
AC ··	BC ··	CB ··	DB ··
or	*or*	*or*	*or*
AD ··	BD ··	CD ··	DC ··

$$=4 \times 3 = 12 \text{ ways}$$

The third space can be filled in these ways:

ABC · *or* ABD ·   BAC · *or* BAD ·
ACB · *or* ACD ·   BCA · *or* BCD ·
ADB · *or* ADC ·   BDA · *or* BDC ·

CAB · *or* CAD ·   DAB · *or* DAC ·
CBA · *or* CBD ·   DBA · *or* DBC ·
CDA · *or* CDB ·   DCA · *or* DCB ·

$$=4 \times 3 \times 2 = 24 \text{ ways.}$$

The fourth space must then be filled with the remaining unused letter:

ABCD	ABDC	BACD	BADC	CABD	CADB	DABC	DACB
ACBD	ACDB	BCAD	BCDA	CBAD	CBDA	DBAC	DBCA
ADBC	ADCB	BDAC	BDCA	CDAB	CDBA	DCAB	DCBA

$$=4 \times 3 \times 2 \times 1 = 24 \text{ ways}$$

This shows us that there are just 24 different ways of entering the four letters in the four spaces. For other numbers of letters and spaces, we can follow the same procedure. For example, if there were seven letters and seven spaces, the number of ways of filling the spaces would be:

$$7 \times 6 \times 5 \times 4 \times 3 \times 2 \times 1 = 5040 \text{ ways}$$

It saves time if, instead of writing '$7 \times 6 \times 5 \times 4 \times 3 \times 2 \times 1$,' we simply write '7!', and call this term 'factorial seven'. The values of the factorial numbers from 0! to 16! are given in Table 82, p. 199.

Summing up in general terms: if there are *n* objects, to be placed in *n* spaces it is possible to do this in *n*! different ways.   (*Formula 1*)

If we had four letters, but only two spaces to fill, we could fill the first space in four ways and the second space in three ways. There would be 12 ways, as shown by the second array on page 130. Mathematically we can write this as:

$$\frac{4 \times 3 \times (2 \times 1)}{(2 \times 1)} = \frac{4!}{2!}$$

Summing up: if there are *n* objects, to be placed in *r* spaces, it is possible to do this in $n!/(n-r)!$ ways.   (*Formula 2*)

This can be taken further. Suppose that we are not interested in the order in which the letters are written, but only in which letters occur together in groups. If this were so, the sequence AB would be equivalent to the sequence BA, and so on. Inspection of the second array on page 130 shows that there are only six different groupings of letters:

   AB   AC   AD   BC   BD   and   CD

Each of these is represented twice, for we have two letters and two spaces, so the number of ways of writing each group is 2! (Formula 1).

This means that if we merely want to know how many ways we can pick out two letters from four (not bothering about their order) we must divide the number of ordered arrangements by 2!, giving us:

$$\frac{4!}{2!2!} = 6 \text{ ways}$$

In general terms, if there are *n* objects to be taken *r* at a time, the number of ways of picking them out is $\dfrac{n!}{n!(n-r)!}$ (*Formula 3*)

This is sometimes written in an abbreviated form as $\begin{pmatrix} n \\ r \end{pmatrix}$.

PROBLEMS

29 In how many ways is it possible to allocate 16 mice to 16 cages, placing one mouse in each cage?

30 In how many ways is it possible to allocate nine experimental treatments to nine plots, allowing one treatment to each plot?

31 An experimenter has six plants and wishes to pick four of these for the experiment; in how many ways can he choose four plants, when the order of choice does not matter?

32 There are 1023 larvae in a dish of water, and five are taken out for examination. In how many ways is it possible to do this, when the order of choice does not matter?

The randomisation test

When we test the data of Table 40, our aim is to show that the two sets of measurements differ—that they are drawn from two different populations. One way to approach this is to first state a null hypothesis, that the two sets of measurements are drawn from the *same* population. If this is so, any difference between the sets is due to random allocation of the measurements to one set or the other set. In effect we made eight trials, and picked out four measurements which we called 'direct viewing', but these four were merely a random selection from eight measurements, all unaffected by viewing conditions. The number of ways in which four measurements can be picked from eight is found by applying Formula 3 above, with $n = 8$ and $r = 4$. This gives:

$$\binom{8}{4} = \frac{8!}{4!4!} = 70 \text{ ways.}$$

If the null hypothesis is true, there are 70 different tables of results, all of which could have been obtained with equal likelihood. For each of these tables it would be possible to work out totals and calculate the difference between times under the two viewing methods. In the observed results of Table 40 (p. 129), the direct viewing total is 48 s and the reflected viewing total is 68 s, giving a difference of 20 s. How many of the 70 possible arrangements of these measurements would produce a reflected viewing total greater than the direct viewing total by 20 s or more? In other words, what are the chances of obtaining *randomly* a result with a difference as great or greater than the one actually observed?

Inspection of the table, and a few trial calculations, show that there is only one other arrangement with a difference as great or greater. This is obtained by exchanging the '15' under direct viewing with the '14' of reflected viewing, and gives a difference of 22 s. Of the 70 possible tables only two (the observed one and the one just described) give a difference as great or greater than the observed difference. We now have to choose between two alternative interpretations of the data:

1. The null hypothesis is **true**. There is no experimental effect. The observed results have arisen at random, the probability of such an occurrence being 2 in 70, or $p = 0.0286$.

2. The null hypothesis is **untrue**. Direct viewing times are on average

shorter than reflected viewing times, the probability of obtaining this result solely by chance being only 0·0286.

Since the probability of random error is so small, this is a risk we would generally accept, and would adopt the second interpretation. The experimental effect is statistically significant at a probability level of 0·0286, or 2·86%.

The randomisation test can also be used when the numbers of observations in each sample are unequal. For example, if there is one extra measurement of, say, 14 s for direct viewing, the numbers of observations in each set become 5 and 4. Formula 3 gives the number of ways of picking from nine observations a set with four observations, leaving a set containing five observations:

$$n = 9 \quad r = 4$$

$$\binom{9}{4} = \frac{9!}{4!5!} = 126 \text{ ways.}$$

To save time, Table 83 has been prepared. For values of n up to 16 it gives the number of ways of picking out two samples; for even n, when samples are equal in size, and for odd n, when samples differ by 1, as above. These values are given in the column headed '2', that is, for two samples drawn from n observations. Also entered in the table is the 'number of ways' multiplied by 0·05, and the reciprocal of the 'number of ways'. Thus for $n = 9$ we find these three entries:

> 126 . . . there are 126 possible ways of choosing the two samples.
>
> 6 . . . if we find that we can set out more than six tables which have a difference of totals as great or greater than that given by the table of observed results, then the probability of obtaining the observed result at random is more than 0·05, and the difference is not significant.
>
> 0·0079365 . . . used when expressing probability as a decimal fraction. For example if $p = 2/126$, we calculate $p = 2 \times 0·0079365 = 0·01588$.

Table 41, which includes the extra value 14 s, shows the only two arrangements of the data which give differences as great or greater than that observed. One of these is the observed table, and the other is got by interchange of a pair of measurements. Two of the values in Table 41 have the same value, 14 s. We say that these values are **tied**. We must handle the tied values as if they were distinct. In reality they probably *are* distinct, for times are recorded to the nearest second, yet one value might really have been 13·77 s and the other, say, 14·16 s.

133

Table 41

Time taken to trace a path (sec); unequal numbers of measurements in samples

Observed results:

Direct viewing	12	11	15	10	14	*Total* = 62	Difference = 6
Reflected viewing	18	16	20	14	—	*Total* = 68	

Interchanged data:

Direct viewing	12	11	14	10	14	*Total* = 61	Difference = 8
Reflected viewing	18	16	20	15	—	*Total* = 69	

For each of the two tables shown in Table 41, there is one more table formed by interchanging the 14's. Altogether there are four tables out of 126 which show differences in the expected direction equal to or greater than that observed. The null hypothesis may be rejected, and the existence of an experimental effect accepted with a probability of $4/126$, or $p = 4 \times 0.0079365 = 0.0317$.

The randomisation test illustrates some of the distinctive features of distribution-free tests. The most important is that we have made no assumptions about the *form* of the distribution of the data. We have worked only with the figures observed, and have not attempted to estimate means or standard deviations of hypothetical populations. The arithmetic involved was very simple, and the levels of probability have been calculated using simple formulae. In general, distribution-free tests are quicker to use, and a calculating machine is hardly necessary. They are the most reliable type of test when data is not normally distributed, and they are particularly suitable for use with small samples. Since they rely less on complex mathematics (such as that surrounding the use of the normal distribution), these tests are more easily understood, and it more readily becomes obvious if one is using them wrongly. The randomisation test has its parallel in the *t*-test, and there are distribution-free tests equivalent to most of the tests based on the normal distribution. Some of the distribution-free tests, for example the *runs test*, have special applications which are not covered by any other tests.

The runs test

If we have four objects of one kind (say, plusses) and four objects of another kind (say, zeros), in how many different sequences may they be arranged? The number of sequences of eight objects is 8! (*see* Formula 1 page 131), but in this example the four plusses are identical, so 1/4! of the 8! sequences are identical on that account, and 1/4!

are identical on account of the four zeros being identical. Thus the number of different sequences is:

$$\frac{8!}{4!4!} = 70$$

These 70 sequences are written out in Fig. 22, where they are grouped into columns according to the number of runs each sequence contains. For example, only 2 sequences have 2 runs, whereas 18 sequences have 4 runs. The columns form a histogram, but this is not a histogram of a set of data; it represents the distribution of the numbers of runs found in all possible sequences of four plusses and four zeros.

Now let us consider an example. A transect was taken across an area of open woodland and the plants found at 32 stations along the transect were recorded. The occurrence of *Glechoma hederacea* (ground ivy) at these stations can be represented by a series of zeros (= absent) and plusses (= present):

+ + + + + + +0000000 +00000 +0 + +000 + + + + +

There are 16 zeros, 16 plusses, and 9 runs. If there is a tendency for *Glechoma* plants to grow in clusters rather than be spread out at random along the transect, we would expect to get long runs of plusses. The total number of runs would be fewer than that expected from a random arrangement. To calculate the likelihood of obtaining a given number of runs, we could write out all possible sequences and derive a histogram as was done in Fig. 22, but since the number of sequences is $32!/16!16!$, which is over 600000000, this is not a practicable method. The numbers of sequences with a given number of runs can be found by calculations involving the application of the formulae we have already used. The application is more complex than in the examples we have already considered, and the results of the calculations have been put in the form of a table, Table 84 (page 201). In our example, $n_1 = n_2 = 16$, and the number of runs $= u = 9$. If we enter the table at $n_2 = n_2 = 16$, and look in the column headed $p \leqslant 0.05$, we find that the critical value for u is 11. This means that of all the possible sequences of 16 plusses and 16 zeros, 5% or fewer contain 11 runs or fewer. We found only 9 runs, so this number is less than the critical value, and would be expected in fewer than 5% of random sequences. In other words, there is good evidence that the null hypothesis of random occurrence of the plants is untrue, and that the plants tend to occur in clusters, or runs. Looking under the column headed $p \leqslant 0.01$, we find the critical value of u is 10; our observed

135

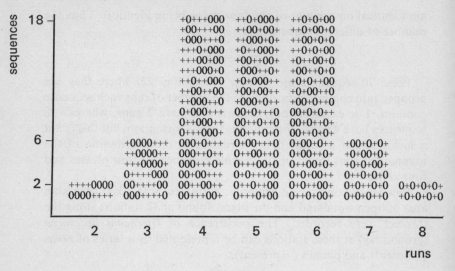

Fig. 22

value of 9 is less than this, so the results are significant at $p = 0.01$ or less (1% level). The column headed $p \leqslant 0.001$ give the critical u as 8. Only 0.1% of sequences have eight runs or fewer, and our observed sequence is not one of these.

Beside each critical value in the table is a probability level printed as a decimal fraction. This is the actual level for the critical value given. As can be seen from Fig. 22, the distribution of runs is discontinuous. It is not possible to find a number of runs corresponding *exactly* to $p = 0.05$, $p = 0.01$ and $p = 0.001$. For example when $n_1 = n_2 = 16$ there is no critical value of u for $p = 0.05$. The probability of getting 12 runs or fewer is 0.052, which is just too high for significance at the 5% level. Therefore the critical value entered in the table is 11 runs or fewer, and for those who want to know the exact significance level of this number of runs, the decimal fraction give the value, $p = 0.023$. As another example take Fig. 22. The probability of three runs or fewer is $8/70 = 0.114$, which is too high for significance, so the table gives the critical value, two runs, which has the probability $2/70 = 0.029$.

The table is calculated for those seqences in which $n_1 = n_2$. In this example the plant was conveniently present at exactly half the number of stations. Provided that the numbers do not differ too greatly, it is possible to use this table for instances when n_1 and n_2 are not equal. If n_1 and n_2 differ by no more than about six, we can enter the

136

table by the row corresponding to the mean of n_1 and n_2. The critical value will be accurate still, though the exact level of probability will not then apply.

When n_1 or n_2 are greater than 20, the table cannot be used, so instead we apply an approximate method, as follows:

Calculate $\quad \mu_u = \dfrac{2n_1n_2}{n_1+n_2} + 1 \quad$ and $\quad \sigma_u^2 = \dfrac{2n_1n_2(2n_1n_2-n_1-n_2)}{(n_1+n_2)^2(n_1+n_2+1)}$

These are the mean number of runs in all possible sequences of this composition, and the variance of the mean number of runs.

Then calculate $\qquad t = \dfrac{\mu_u - u}{\sigma_u}$

Compare this value of t with those in Table 74 (p. 192) for an infinite no. of degrees of freedom. The t-table is calculated for deviations *both above and below* the mean, but in this test we are concerned only with deviations *below* the mean. Only a number of runs *less* than that expected can cause us to reject the null hypothesis. For this reason, we halve the probability levels quoted in the table.

For example, on a longer transect, with 152 stations, $n_1 = 36$, $n_2 = 116$ and $u = 19$. By calculation we find that $\mu_u = 55$, and $t = 8.2$. Consulting Table 74 we find that p is less than 0.001. The value of t relates to sequences of 19 or fewer runs (deviating 36 runs or more below the mean) *and* to sequences of 91 or more runs (deviating 36 runs or more above the mean). There is an equal number of sequences in either category, and we discount those with excess runs, so we read the probability level as $p = 0.0005$.

Other applications of the runs test

The original data of the experiment referred to on page 51 are given in Table 42.

Table 42

Bud lengths (mm)

IAA	26	28	37	41	57	58	67	78	80	121	122
Control	15	30	30	65	91	103	132	135	137	141	198

If we write out the data as a single sequence, and below each measurement write an i or a c, according to whether it comes from the IAA treatment or the control we get Table 43.

The null hypothesis is that IAA treatment has no effect, that all measurements are from a single population, and that in effect the labels i and c are being allocated at random. According to the null hypothesis, any one of the possible sequences of i's and c's is equally

137

Table 43

Bud lengths (mm); data as a labelled sequence

15 26 28 30 30 37 41 57 58 65 67 78 80 91 103 121 122 132 135 137 141 198

c i i c c i i i i i c i i i c c i i c c c c c

$$n_2 = n_2 = 11 \qquad u = 9$$

likely to occur. If the samples are drawn from different populations, because of the effect of IAA treatment, the measurements from one sample will have the tendency to cluster together, and there will be fewer runs. According to Table 84, we would require seven runs or fewer before accepting this result as significant at the 5 % level, so no experimental effect has been demonstrated.

Had too few runs been indicated, this would have told us that measurements tended to cluster, and that the populations differed. Clustering can be caused by differences of means, or differences of standard deviation, or both differences acting together, so that though we can demonstrate that two populations differ, we cannot assess the nature of the difference by using this test.

The problem of tied values has not arisen in this example, for the only tied values both belong to the same treatment. If two tying values occur, one in each treatment, we resolve the first such tie so that *c* precedes *i*, and the next tie so that *i* precedes *c*, and so on. Odd ties are resolved at random, for instance by tossing a coin. If there is a large proportion of tied values, the test is not reliable.

The runs test may also be used to detect a trend in a series of observations. Consider the data for *Agrostis* in Table 37 (p. 119). The percentages appear to decrease along the transect from sample 1 to sample 16. To test this effect we first find the median percentage cover value, which lies between 40% and 43%. Next we list the values as a sequence in order, labelling each according to whether it is above the median (*a*) or below the median (*b*), as in Table 44.

Table 44

Cover values (%): data as a labelled sequence

Sample no.	1	2	3	4	5	6	7	8	9	10	11	12	13	14	15	16
Cover value	60	52	63	51	43	52	38	56	44	23	40	23	32	31	14	20
a or b	a	a	a	a	a	a	b	a	a	b	b	b	b	b	b	b

$$n_1 = n_2 = 8 \qquad u = 4$$

Table 84 indicates that this result shows a significant departure from randomness at the 1 % level ($p = 0.009$). The values below the median are tending to cluster at one end of the sequence and the values above the median are tending to cluster at the other end. This indicates a significant trend of cover value along the transect.

This test can also be applied to other series of measurements, such as the numbers of aquatic organisms recorded in daily samplings of pond water, or the weight of a growing animal recorded at daily or weekly intervals. When testing for trend, some values may tie with the median. If so, these should be allocated at random, half above the median (labelled *a*) and half below it (labelled *b*). As before, too many ties will reduce the reliability of the test.

Where there are more than two samples

This test may be used in place of a 1-factor analysis of variance. It requires less labour in the calculation, and has the advantage that the number of replicates need not be the same for each treatment. This is specially useful when there are missing values. Another advantage is that the test can be used either with measurement data or with **ranked data.** Sometimes an experimental effect cannot be measured in any precise way, yet the subjects of the experiment can be ranked in order by an experienced observer. For instance, a series of ten bacterial cultures in broth could be assessed for turbidity, ranking them from 1 (clear or almost clear) to 10 (the most turbid). No measuring instrument would be needed, for they could be judged by eye, yet this crude technique will often give results sufficiently precise at a considerable saving of time and expense. Sometimes the effects to be assessed are actually unmeasurable; the extent of damage of a series of plants by a fungus disease is not easily assessed quantitatively but an observer could rank the plants in order of severity of attack by the fungus.

To illustrate the working of this test, we will use the same data as for the demonstration of the 1-factor analysis of variance, from Table 12, p. 56. In Table 45 this data is recorded after ranking the measurements from 1 (lowest weight) to 16 (highest weight). Tied values are all allocated the mid-rank. For example, the 13's in the 'Both' column occupy ranks 2, 3, and 4, so they are all given the mid-rank, rank 3. Where an even number of values are tied, half-ranks are used; for example if two values tied for ranks 7 and 8, they would both be allocated mid-rank, rank $7\frac{1}{2}$.

The null hypothesis is that there is no experimental effect and that the arrangement of ranks within the table is a random one. There are 16 ranks, so there are 16! ways in which they could have been entered in the table (Formula 1, p. 131). But we are not concerned with the

Table 45

Water lost from Cherry Laurel leaves in three days (ranked data)

| Leaf replicate no. | Surface covered with jelly | | | |
	Neither	Top	Bottom	Both
1	14	10	5	3
2	15	11	7	1
3	16	9	8	3
4	13	12	6	3
Rank totals	58	42	26	10

order of arrangement *within* columns, and in a given table each column can be formed in 4! ways. So *for each column* we must divide 16! by 4!. Further, we are planning to demonstrate simply an overall effect of treatment, not a set of comparisons between one specified treatment and another. Tables which can be formed simply by interchanging the four complete columns of figures are identical for the purposes of this analysis, so we must put another 4! into the divisor. The number of distinct tables is

$$\frac{16!}{4!4!4!4!} = 2627625 \text{ tables}$$

This figure can be obtained by consulting Table 83 (p. 200).

The next step is to determine how many of these show treatment effects as great or greater than those shown by the table of observed results. The criterion for assessing the effect of treatment is proportional to the sum of the squares of the rank totals, so our task is to find out how many tables have sums of squares of rank totals as great or greater than that of Table 45. Sometimes, especially when experimental effects are slight, there may be many such tables. The work of discovering them all may become so tedious that it is preferable to use the analysis of variance. In this example there is no difficulty, for inspection of the table shows that there is no overlap between the columns. It is not possible to increase any column total except by interchanging a rank value with a column which already has a greater rank total; such as interchange will *reduce* the sum of squares. So Table 45 shows the experimental effects better than any other table which could be produced, and the probability of obtaining this table at random is 1 in 2627625, or $p = 0.00000038$. The null hypothesis must be rejected, and the existence of an experimental effect accepted with an extremely high level of probability. This is confirmed by the results of the analysis of variance (pp. 163–5).

In Table 46 the data of Problem 15 (p. 62) are set out as ranks, with one value missing (suppose there was a heavy rainstorm before this last measurement could be taken).

Table 46

Soil temperatures (ranked data)

	Sunny garden bed	Under evergreen shade	Shade, under leaf litter
	13	7	2
	14	7	2
	10	7	7
	11	7	2
	12	4	
Rank totals	60	32	15

There are two sets of ties in this data. Three measurements of 9° C tie for ranks 1, 2, and 3, so they are given mid-rank, rank 2; five measurements tie for ranks 5 to 9, so they are given rank 7. Fortunately, the occurrence of a large number of ties does not affect the validity of this test.

There are 14! ways of entering the ranks in this table, but we are not concerned with different arrangements within the columns, so we divide 14! by 5!5!4!, the product of the factorials of the numbers of observations in each column. There are only two columns which can be interchanged as a whole, since the third contains fewer ranks, and so there are 2! ways of rearranging whole columns without producing distinct tables. Thus the number of distinct tables is

$$\frac{14!}{5!5!4!2!} = 126126 \qquad \text{(see Table 83, p. 200)}$$

The sum of squares of rank totals of Table 46 is

$$60^2 + 32^2 + 15^2 = 3600 + 1024 + 225 = 4849$$

There is only one way in which this sum can be increased, which is by interchanging the two ranks printed in bold type. The sum of squares then becomes

$$60^2 + 35^2 + 12^2 = 3600 + 1225 + 144 = 4969$$

The probability of obtaining at random a table which shows experimental effects as great or greater than those shown by Table 46 is 2 in 126126, or $p = 0.0000158$. The null hypothesis can be rejected, and the effect of situation on soil temperature accepted as demonstrated with a very high degree of probability.

141

Another rank test

The Wilcoxon rank test is used for comparing two samples of measurement data or rank data. As an example, we can take the earlier observations recorded when the data for Problem 2 (p. 14) was being collected. The data is first listed in two columns, and then ranks are allocated, beginning with rank 1 for the lowest measurement. The rank total is calculated for the column with the least number of observations. The results of this calculation are seen in Table 47.

Tied values have been dealt with by assigning them the mid-rank, as described for the previous test. The conjugate total is the rank total which would have been obtained if the ranks had been allocated in the reverse order, beginning with rank 1 for the highest measurement. The equation saves re-ranking.

The final step is to compare either the column total, T, or the conjugate total, T', whichever is the lesser, with the critical values given in Tables 85 and 86. If T or T' are less than the critical value, this indicates that the null hypothesis should be rejected. The tables have been prepared by calculating how many possible arrangements of ranks can be written out for each pair of values of n_1 and n_2, and then dis-

Table 47

Cotyledon widths (mm)

| | No fertiliser | | Fertiliser provided | |
	mm	rank	mm	rank
	18	$14\frac{1}{2}$	29	25
	19	18	17	$11\frac{1}{2}$
	12	2	21	$22\frac{1}{2}$
	16	$8\frac{1}{2}$	21	$22\frac{1}{2}$
	14	5	30	26
	17	$11\frac{1}{2}$	19	18
	13	3	17	$11\frac{1}{2}$
	16	$8\frac{1}{2}$	17	$11\frac{1}{2}$
	15	7	19	18
	14	5	11	1
T = Rank total =		83	18	$14\frac{1}{2}$
			20	21
			19	18
			27	24
			14	5

$$n_1 = 10 \qquad n_2 = 16$$
$$T' = Conjugate\ total = n_1\ (n_1 + n_2 + 1) - T = 10(10 + 16 + 1) - 83 = 187$$

covering the critical value below which 5% of totals or conjugate totals will fall. The procedure for calculating the table is too elaborate to be described here, but it follows the same principles as in many of the previously described tests.

In the example given above, the total is less than the conjugate total, and its value is 83. In Table 85, for $n_1 = 10$ and $n_2 = 16$ the critical value is 97, so we can reject the null hypothesis. The data indicate an significant experimental effect. As in the runs test (pp. 134–9), the test simply indicates *a difference*; it does not tell us whether this is a difference of mean, of standard deviation or of both.

There are two points not covered by the example above. If $n_1 = n_2$, total both columns and compare the lesser total with the critical value of the table. If there are tied values which bring a rank total to an odd half, raise the total to the next integer.

Rank correlation

An experiment was conducted to test the effects of the concentration of phenol on the amount of growth of bacteria.[1] When the culture tubes were examined, the growth was found to consist partly of bacteria in suspension and partly of irregular colonies, producing a flocculence which it was impossible to measure in a meaningful way. Accordingly, the amount of growth was assessed visually and the culture tubes ranked. This experiment was performed independently by two groups, and their rankings and the calculations required for the test are shown in Table 48. Growth was least at high and at low concentrations of phenol, but the concentration showing the greatest growth was not the same in both sets of results. The coefficient r_s is Spearman's coefficient of rank correlation. It ranges in value from +1, when there is complete agreement between two sets of ranks, to –1, when there is complete disagreement, that is when the ranking of one group is in exactly the opposite order. It is possible to set out all pairs of ranking orders and to calculate r_s for all these arrangements. Then the critical coefficients for the 1% and 5% most concordant arrangements of ranks can be found and tabulated. Table 87 (p. 204) shows the results of such calculations. The coefficient calculated from the data above exceeds the tabulated value for $n = 7$, $p = 0.05$, so this data is more concordant than would be expected in a purely random arrangement of ranks. The agreement or correlation between the two sets of data is statistically significant.

[1] See *Nuffield Biology Text II*, Longman/Penguin Books.

143

Table 48

Growth of bacteria (ranked values)

		Ranks		
Concentration	Group 1	Group 2	Deviation	d^2
of phenol (%)	(x)	(y)	$(d = x - y)$	
0·2	1	2	−1	1
0·1	2	1	1	1
0·05	3	3	0	0
0·025	5	7	−2	4
0·0125	7	6	1	1
0·00625	6	5	1	1
0 (*control*)	4	4	0	0
				$\Sigma d^2 = 8$

$$r_s = 1 - \frac{6\Sigma d^2}{n(n^2 - 1)} = 1 - \frac{6 \times 8}{7 \times 48} = 1 - \frac{1}{7} = 0.86$$

The treatment of tied ranks is different in this test. To give them mid-rank values tends to bias the correlation coefficient, especially when n is small, and a correction must be applied to the calculation. A better procedure is to resolve the tied ranks in the manner least conducive to the rejection of the null hypothesis. In other words, we seek to make r_s smaller by making d^2 bigger, and thus err on the safe side. Then we shall not state that correlation is significant when it is not (though we might occasionally say it is not significant when it is). For example, if in the phenol experiment, Group 1 had been unable to see any difference between their 0·2% and 0·1% tubes, they would tie for ranks 1 and 2. This could be resolved in two ways:

x	y	d	d^2		x	y	d	d^2
0·2	1	2	−1	1 *or*	2	2	0	0
0·1	2	1	1	1	1	1	0	0

The first way of resolving contributes 2 to d^2, whereas the second way contributes nothing. The tie must be resolved in the first way.

A test for choice-chamber experiments

In many experiments on animal behaviour an individual is put into a choice-chamber apparatus in which it has the choice of entering one or other of two chambers. If these chambers are equally large and accessible, and there is no experimental effect, the individual is equally likely to enter either chamber, and the probability can be calculated of finding that from, a set of, say, ten individuals, as many as eight will enter chamber A and only two will enter chamber

B. The probability that nine will enter one chamber and only one will enter the other is 0·021, so that if such behaviour is shown it is more reasonable to suppose that the individuals are not behaving randomly, but are exercising some definite choice, presumably because of differences between the two chambers. The apparatus is usually designed so that there is only one difference between chambers; one chamber may be light, the other dark, or one may be damp and the other dry, and so on. The application of this test rests on the assumption that all individuals are unaffected by all other individuals. For instance, if all ten were tested simultaneously, an individual might not enter a chamber simply because it already contained another individual. If the individuals were tested successively in the same apparatus, it is possible that a scent left behind by an individual tested earlier might influence subsequent trials. Such possibilities have to be kept in mind and, if necessary, eliminated by suitable techniques. If trials are truly independent, and chambers equally likely to be chosen (in the absence of experimental effect), we can use Table 88 (p. 204) to assess the significance of results. Then any result which is as extreme or more extreme than the distribution shown in the table is significant at the level indicated in by the decimal fractions.

PROBLEMS

33 Fourteen batches of barley seed were counted out, each batch containing 50 seeds. Seven of the batches were placed on moist filter paper for six days, during which time they began to germinate. The other seven batches were left dry. After six days all batches were dried to constant weight in an oven, and the dry weights measured. The dry weights (g) were:

Germinated seeds:	1·49	1·51	1·40	1·38	1·32	1·51	1·35
Ungerminated seeds:	1·48	1·62	1·53	1·56	1·53	1·55	1·52

Use the randomisation test to decide whether there is a significant loss of dry weight in the germinated seeds.

34 On the transect referred to on p. 134, the distribution of *Urtica dioica* (stinging nettle) was:

$$0+ +000000000000 + + + + + + + + + + +00 + +000 +0 +$$
$$(0 = \text{absent}, \ + = \text{present})$$

Use the runs test to decide if this distribution shows that nettle plants tend to occur in clusters.

35 On a longer transect, *Urtica dioica* was present at 67 stations, and absent from 85 stations; there were 37 runs. Use the runs test to decide if this distribution gives evidence of clustering.

36 When part of the data for Problem *6* (p. 15) had been gathered, it read as follows:

Small-flowered   15   9 13   8 13   9   7 10 11   9

Normal-flowered 15 19 14 20 17 12 21 17 14 16 12 19 12 17

Use the runs test to discover the significance of the difference between the two samples.

37 Use the runs test to analyse the data of Problem *33*.

38 The percentage cover values for *Juncus*, given in Table 37, (p. 119) show an apparent trend from one end of the transect to the other. Use the runs test to investigate this.

39 Use the rank test for more than two samples (p. 139) to analyse the soil water content data of Problem *16*, (p. 62). Repeat with an extra value 4% in the centre column.

40 Using the data of Problem *17*, (p. 63), test the significance of the variation of oxygen content with depth. Ignore differences between sites, considering the sites as replicates.

41 Test the data of Problem *36*, using the Wilcoxon rank test.

42 Test the data of Table 42 (p. 137) using the Wilcoxon rank test.

43 In a bacteriological investigation, subjects washed and dried their hands using several different procedures, and then wiped their fingers on the surface of sterile nutrient agar medium. After the culture dishes had been incubated, they were ranked according to the amount of bacterial growth present. It was not possible to give a numerical estimate of contamination by simply counting the colonies, because in some instances colonies were numerous (well over 1000), closely packed and overlapping. In other instances there were several different kinds of colony on a dish. The experiment was performed by two groups, and the results obtained were:

	Ranking	
Treatment	*Group 1*	*Group 2*
Hands not washed	6	5
Hands washed with soap, dried on communal towel	5	6
Hands washed with soap, dried on clean paper tissue	4	4
Hands washed with soap, dried in hot air	3	2½
Hands washed with antiseptic washing compound, dried in hot air	2	2½
Unopened control dish	1	1

Calculate Spearman's coefficient of rank correlation for this data, and assess the significance of the correlation between the ranking in the two groups.

12 Planning the analysis

Earlier chapters have dealt in detail with a selection of the most useful statistical tests. This chapter summarises the tests, and will help in the choice of test most suitable for a given set of results.

All the techniques are listed in Table 50 which, in conjunction with the explanation of mathematical symbols and equations given in Table 49, will serve as a summary guide to the whole book. The

Table 49

Symbols and equations

Symbol	Interpretation	Equations
Σ	summation	
x, y	variables	
n	number of observations in a sample	
$\bar{x}, \bar{y}$	arithmetic means of a sample	$\bar{x} = \Sigma x/n$
f	frequency of a given value of the variable	$\Sigma f = n$
Σd^2	sum of squares of deviations from a sample mean; $\Sigma d^2{}_x, \Sigma d^2{}_y$ distinguish between variables, x and y	$\Sigma d^2 = \Sigma(x - \bar{x})^2$ OR $\Sigma d^2{}_a = \Sigma x^2 - (\Sigma x)^2/n$
$\Sigma d_x d_y$	Sum of products of deviations in x and in y	$\Sigma d_x d_y = \Sigma xy - \Sigma x \Sigma y/n$
s^2	variance of a sample	$s^2 = \Sigma d^2/n$
s	standard deviation of a sample	
μ	mean of a population	best estimate is $\bar{x}$
σ^2	variance of a population	best estimate is $\Sigma d^2/(n-1)$
σ	standard deviation of a population	
σ_n	standard deviation of the mean of a sample of n observations	$\sigma_n = \sigma/\sqrt{n}$
σ_d	standard deviation of a difference between 2 means	$\sigma_d = \sqrt{\left(\dfrac{\sigma_1^2}{n_1} + \dfrac{\sigma_2^2}{n_2}\right)}$
σ_σ	standard deviation of a standard deviation	$\sigma_\sigma = \sigma/\sqrt{(2n)}$
t	a normally distributed deviate, expressed in units of standard deviation	
F	variance ratio	$F = \dfrac{\text{mean square of sample means}}{\text{residual mean square}}$
r	coefficient of correlation	$r = \dfrac{\Sigma d_x d_y}{\sqrt{(\Sigma d_x^2 . \Sigma d_y^2)}}$

Symbol	Interpretation	Equation
b	gradient of regression line	*For y on x* $$b = \Sigma d_x d_y / \Sigma d_x^2$$ *For x on y* $$b = \Sigma d_x d_y / \Sigma d_y^2$$
	Regression equations	*For y on x* $\quad y = \bar{y} + b(x - \bar{x})$ *For x on y* $\quad x = \bar{x} + b(y - \bar{y})$
χ^2	chi-squared	$$\chi^2 = \Sigma \frac{(O - E)^2}{E}$$
O	observed values, in contingency tables	
E	expected values, in contingency tables	
p	probability; $O < p < 1$	
l	limits of a statistical quantity, to a given level of significance	$l = \pm t \, \sigma$
u	number of columns ⎫ in analysis of variance	
v	number of rows ⎭	$uv = n$
u	number of runs, in runs test	
T, T'	rank total and conjugate rank total, in Wilcoxon rank test	
r_s	Spearman's coefficient of rank correlation	

degree of condensation of Table 50 makes some amplification necessary.

The data for analysis will fall under one of these headings:

1. *Measurement data:* examples; heights, weights, numbers of flowers, percentage cover. For certain distribution-free tests it is sometimes convenient to convert measurement data to ranked data (see below).

2. *Enumeration data:* the numbers of individuals falling into two or more distinct categories. For example, the numbers of red-flowered plants, and the numbers of white-flowered plants; the numbers of people with blue eyes, and the numbers of people with non-blue eyes; the numbers of mice caught in specified traps; the numbers of snails bearing each kind of distinct camouflage pattern. Sometimes the distinction depends on one criterion (e.g. flower colour) with several categories (red, pink, white), sometimes there may be more than one criterion, as when a quadrat may be classified according to whether or not it contains *Agrostis* **and** whether or not it contains *Juncus* — two criteria (occurrence of the two species) each with two categories (present, or absent).

3. *Ranked data:* obtained when precise measurement is difficult, or impossible. Examples: visual estimations of the amount of growth in a broth culture, assessment of the extent of chlorosis in a set of mineral-deficient plants.

4. *Other data*: examples: the results of choice-chamber experiments, clustering of species along transects.

When dealing with measurement data, there is a choice to be made between using tests based on the normal distribution and the distribution-free tests. In general the choice should be based on what is known, or not known, about the likely distribution of the variable or variables. With small samples, distribution-free tests are often more appropriate, and usually more reliable. With large samples it is easier to assess whether or not the data are normally distributed and choose the appropriate test. However, the randomisation test and the three-sample rank test become too tedious when samples are large. Another limitation is that the runs test cannot be used to compare samples when a large number of tied values occur. Among the many tests based on the normal distribution the choice is made according to the number of variables and the number of samples to be compared.

With a single variable, measured for only one sample, there is no basis for making a comparison, but it is possible to make simple statistical statements about the sample, giving its mean and standard deviation. From these, one can estimate the mean and standard deviation of the population. One may also calculate statistical limits for the estimates, using the distribution of t. When there are two samples, one can perform the operations listed above on each sample independently. Population estimates may then be compared, using the t-test for comparison of the population means (μ_1, μ_2), and the F distribution for comparing population variances (σ_1, σ_2). The comparison of two samples is the essence of much experimental and observational work in biology; one may aim to show that the populations are significantly different, and the differences are related to the experimental treatment, or one may hope to show that there is no evidence of difference, that the two samples are drawn from a single population. Table 50 includes the special technique used in Example 2 (p. 99) where one forms pairs of corresponding observations, and treats the differences between members of each pair as a single sample. With three or more samples for comparison, the simple t-test will frequently fail to give reliable results (p. 57), so a better test is the analysis of variance, which has additional advantages. One can calculate the means ($\bar{x}_1$, $\bar{x}_2$, $\bar{x}_3$, etc.) for every treatment, establish whether there is an all-over significant treatment effect, and then examine the differences between pairs of means in more detail. When comparing two samples, it does not matter if they do not contain an equal number of observations. For the usual analysis of variance the

149

Table 50

Statistical techniques		Data available	
MEASUREMENT DATA	1 variable (x)	1 sample	
		2 samples	if observations pairable, treat differences as a single sample (99
			if observations *not* pairable
		3 + samples	1 factor, with replication
			2 factors, without replication
			with replication
			3 + factors latin square design
			other designs
	2 variables (x, y)	analyse each variable singly, as above	
		1 sample	
		2 + samples	analyse each sample singly, as abo
			compare samples
	3 variables	analyse each variable singly, as above	
		analyse pairs of variables, as above	
		analyse variables simultaneously	
ENUMERATION DATA	1 criterion	2 + categories	
	2 criteria	2 categories	
	2 + criteria	2 + categories	
RANKED DATA	1 variable	2 samples	
		2 + samples	
	2 variables	1 sample	
OTHER DATA	Sequences of + and 0 etc.		
	Choice-chamber data		

Where s and σ are given, also read s^2 and σ^2
Italic figures refer to explanations of tests.

Techniques of analysis available	Population estimates and tests of significance	Distributions used
...lculate $\bar{x}$, s (*16*, **154**, *23*, **157**, **179**, **181**)	μ, σ, σ_n, σ_σ (*28*, **157**, **162**, **179**, **181**) limits of μ, σ (*37*, *41*, **160**, **162**)	t
...lculate $\bar{x}_1$, $\bar{x}_2$, s_1, s_2 (*16*, **154**, *23*, **157**, **179**, **181**)	μ_1, μ_2, σ_1, σ_2, σ_n, σ_σ, σ_d (*28*, *48*, **157**, **161**, **162**, **179**, **181**) limits of μ, σ, σ_d (*41*, *52*, **160**, **162**) compare σ_1^2 and σ_2^2 (*53*, **162**)	t F
...^{F} randomization test (**132**) ...^{F} runs test (**137**)	significance of sample differences	—
...^{F} test for 3 + samples (**139**) ...alysis of variance (*57*, **163**, **184**) ...culate x_1, x_2, x_3 etc. ...culation of missing values (**173**)	significance of sample differences residual σ^2 μ_1, μ_2, μ_3 etc. significance of treatment effects limits of μ sig. diff. of pairs of means (*105*)	— F t t
...alysis of variance (*61*, **165**)		
...alysis of variance (*93*, **171**)	+ significance of interaction	F
...alysis of variance (*89*, **170**)		
...ltifactorial analysis of variance*		
...efficient of correlation, r (*65*, **175**, **187**)	significance and sign of correlation	r
...gression equations, x, y (*67*, **176**, **187**)	μ for x and y, b	
...^{F} runs test for trend (**138**)	significance of trend	—
...efficients of correlation, r_1, r_2 ...gression equations, x_1, x_2, y_1, y_2	comparison of r_1, r_2* comparison of b_1, b_2*	
...ltiple regression*		
...test on $1 \times n$ contingency table (*73*, **176**)	significance of departure from expectation, or significance of association	χ^2
...test on 2×2 contingency table (*76*, **177**)		
...test on $n \times n$ contingency table*		
...F Wilcoxon's rank test (**142**)	significance of sample differences	—
...F test for 3 + samples (**139**) ...F Spearman's test (**143**)	significance of sample differences significance of correlation of ranking	— —
...F runs test for trend (**138**)	significance of trend	—
...F runs test for clustering (**134**)	significance of clustering	—
...F choice-chamber test (**144**)	significance of departure from random choice	—

Bold figures refer to practical instructions for tests
* denotes techniques not found in this book
DF denotes a distribution-free test

numbers must be equal. Missing values, due to accidents or gross errors in the experiment, may be calculated by the equations given on p. 173, and one can then proceed with the analysis in the usual way, having lost one or two degrees of freedom. When there are two factors and the treatments are replicated it is possible to gain information about interaction between these factors. If it can safely be assumed or shown by preliminary experiment that no interaction exists, an experiment based on a latin square can be used for testing the effects of three factors with a relatively small number of observations. In advanced texts the reader will find other ways of using the analysis of variance, to cover experiments in which the number of observations in each treatment are not equal, and designs involving three or more factors which allow for the estimation of interaction.

When there are two variables for each individual of the sample, each variable may be examined separately, using the methods already dealt with above. To determine if there is a relationship between the two variables, one calculates the coefficient of correlation and, if this is significant, the regression equations. If there are two such samples, each may be treated separately, and their coefficients of correlation, and regression equations found. Statistical tests have been devised to establish the significance of the difference between coefficients of correlation of two samples, and between the values of b in the regression equations of two samples, but these tests are not included in this book.

If the data have three or more variables, each may be considered separately, or they may be taken two at a time, and examined for correlation, as in the paragraph above. By the more complex technique of multiple regression analysis, the simultaneous correlations between three or more variables may be studied, but this is beyond the scope of the methods described here.

For dealing with the simpler types of enumeration data, one can use the $1 \times n$ or 2×2 contingency tables, testing the probability of given deviations by using the χ^2 distribution. For contingency tables of higher order, extensions of the usual methods may be used, as described in more advanced texts.

Having decided which technique is the most appropriate to the problem in hand, there is one further preliminary decision to be made. Do the data need transformation? The most commonly used transformations are listed in Table 51. If the data are extensive and show satisfactorily significant results as they stand, it is unlikely that the extra labour of transforming them will be worth while. If the data

Table 51
Transformations

Transformation	When to use it	Example
$\log x$	Examination of sample data shows s is proportional to x	p. 106
ARCSIN (Table 81) (a) θ, when $\sin \theta = \sqrt{x}$	x is a proportion, with upper limit 1, and lower limit 0	p. 117
(b) θ, when $\sin \theta = \sqrt{(x/100)}$	x is a percentage, with upper limit 100, and lower limit 0	p. 118
$\sqrt{x}$	x is discontinuous, and has a lower limit 0	Could have been used on p. 95
$\sqrt{(x+0\cdot5)}$	As above, when any of the observations are less than 10	Nos. of galls on leaves, or caterpillars on plants, or colonies in a petri-dish culture.

shows differences which are marginally significant, and *provided that they come into one of the categories described in Table 51*, they should be transformed before analysis. In certain experiments there may be special reasons for transformation, as in Example 7, (p. 123). Transformations of this kind are many and varied: they are not listed in the table, since they depend upon the theoretical basis of the experiment. The aim of a transformation is to convert non-normally distributed data into normally distributed data. If it seems doubtful that this can be achieved by transformation a distribution-free test should be considered as an alternative. This will often be considerably less laborious, yet just as reliable.

F

13 Practical statistics

Statistical calculations require absolute accuracy, and all working must be checked for correctness of computation. For these reasons, a calculating machine is of very great help, and special methods for use with a machine are given in Chapter 14. Slide-rules, tables of logarithms, and tables of squares and square roots are not of much use, since their four-figure accuracy is generally insufficient. To use these tables may often lead to serious accumulated error by the end of an analysis. All calculations must therefore be worked in full, except where there are heavy lines in the margin of the text, when a slide-rule or four figure tables may be used. Many analyses require the squaring of numbers and a special table (Table 72, pp. 189 – 91) of squares of numbers from 0 to 999 has been prepared with complete, six-figure accuracy. This may be used at *any* stage of analysis. The entries in this table have up to three figures; if the data of the experiment have more than three figures, these may often be rounded off to three figures without any real loss of accuracy, thus simplifying the calculations and making it possible to use Table 72. A table of square roots of integers from 0 to 100 (Table 73, p. 191) has been included. It has five-figure accuracy, and may be used to find $\sqrt{n}$ or $\sqrt{(2n)}$ when calculating σ_n or σ_σ.

To find any particular section of this chapter, refer to the index, or to Table 50, (pp. 149 – 50).

Means

1. WHEN THERE ARE ONLY A FEW OBSERVATIONS
Use the straightforward method of summing the observations, and dividing by the number of observations. Repeat, to check working.

Example: To find the mean of 3·54, 3·51, 3·58, 3·49, and 3·55

$$\Sigma x = 3·54 + 3·51 + 3·58 + 3·49 + 3·55 = 17·67$$

$$\text{Mean, } \bar{x} = \frac{\Sigma x}{n} = \frac{17·67}{5} = 3·53 \text{ (to 2 decimal places)}$$

2. WHEN THERE ARE MANY OBSERVATIONS

This method is designed to make calculation simpler, and to include a check on the correctness of working. With a calculating machine, use instead the method described on p. 179.

Proceed as follows (see the two worked examples in Tables 52 and 53):

Table 52

Calculation of mean: the numbers of seeds in ripe pods of Laburnum

Observations are not grouped; class interval, $c=1$; working mean, $w=3$.

No. of seeds per pod x	No. of pods f	D	fD	$(D+1)$	$f(D+1)$
1	27	-2	-54	-1	-27
2	32	-1	-32	0	0
3	21	0	0	1	21
4	15	1	15	2	30
5	13	2	26	3	39
6	9	3	27	4	36
7	4	4	16	5	20
8	1	5	5	6	6
	$\Sigma f = 122$		$\Sigma fD =$ $89-86=3$		$\Sigma f(D+1) =$ $152-27=125$

CHECK: $\Sigma f(D+1) - \Sigma fD = 125 - 3 = 122 = \Sigma f$ Working is correct so far.

CALCULATION OF MEAN:
$$\bar{x} = w + \frac{c\Sigma fD}{\Sigma f} = 3 + \frac{1 \times 3}{122} = 3 + 0.0246 = 3.0246 \text{ seeds.}$$

Table 53 Calculation of mean: the number of flowers on antirrhinum plants (data from Table 5)

Observations grouped; class interval, $c=20$; working mean is the middle value of the modal class, $60-79$, so $w=69.5$.

No. of flowers per plant x	No. of plants f	D	fD	$(D+1)$	$f(D+1)$
20– 39	6	-2	-12	-1	-6
40– 59	10	-1	-10	0	0
60– 79	12	0	0	1	12
80– 99	7	1	7	2	14
100–119	5	2	10	3	15
120–139	1	3	3	4	4

155

Table 53 (*continued*)

No. of flowers per plant x	No. of plants f	D	fD	(D+1)	f(D+1)
140–159	1	4	4	5	5
160–179	1	5	5	6	6
	$\Sigma f=43$		$\Sigma fD=$ $29-22=7$		$\Sigma f(D+1)=$ $56-6=50$

CHECK: $\Sigma f(D+1)-\Sigma fD=50-7=43=\Sigma f$ Working is correct so far.

CALCULATION OF MEAN:

$$\bar{x}=w+\frac{c\Sigma fD}{\Sigma f}=69{\cdot}5+\frac{20\times7}{43}=69{\cdot}5+3{\cdot}256=72{\cdot}756 \text{ flowers.}$$

(a) Set out a table of six columns, headed as in Tables 52 and 53.

(b) Enter the observations in the first two columns, as a frequency distribution. If no value of the variable occurs more than once, f is 1 for all classes, and the second, fourth and sixth columns may be omitted. The class interval of the frequency distribution may be 1, with ungrouped data (Table 52) or may take other values (Table 53), but all class intervals must be the same.

(c) Sum the second column to find $\Sigma f=n$, the total number of observations.

(d) By examination of the first two columns, estimate which class contains the mean; this will usually be the modal class. Take the middle value of this class as the **working mean**, w.

(e) In the third column, enter values of D, which is the deviation of each class from the class containing the working mean. This deviation is expressed in class intervals. Therefore enter 0 at the class containing the working mean, then proceed up the column from the zero, and enter -1, -2, -3, and so on; next proceed down the column from the zero and enter 1, 2, 3, and so on to the bottom of the column.

(f) Calculate the values of fD, and enter these in the fourth column.

(g) Sum the fourth column, to find ΣfD.

(h) As a check, enter $(D+1)$ in the fifth column, calculate $f(D+1)$, and enter these values in the sixth column. Sum the sixth column to get $\Sigma f(D+1)$.

Check that: $\Sigma f(D+1)-\Sigma fD=\Sigma f$. If this is so, the calculations are correct so far. If not, examine the calculations for error.

(i) Calculate the mean from the equation:

$$\bar{x}=w+\frac{c\Sigma fD}{\Sigma f}, \text{ where } c \text{ is the class interval.}$$

This equation corrects the working mean by adding $c\Sigma fD/\Sigma f$ to it. $\Sigma fD/\Sigma f$ is the mean of deviations from the working mean, and this fraction when multiplied by c restores the units in which deviations are measured from units of class interval to the original units of the data. If the working mean is greater than $\bar{x}$, $c\Sigma fD/\Sigma f$ will be negative.

For this step of the calculation logarithms or a slide rule may be used, provided that no more than 3 figures are required in the answer.

Variance and Standard Deviation

In these methods one first calculates the variance (s^2, or σ^2), from which the standard deviation (s, or σ) may be found by taking the square root.

1 WHEN THERE ARE ONLY A FEW OBSERVATIONS:

(a) Tabulate x and x^2; find Σx and Σx^2 by summing the columns of the table; apply the equation $\Sigma d^2 = \Sigma x^2 - (\Sigma x)^2/n$, where n is the number of observations. This gives the value of Σd^2, which is used below.

(b) To find variance and standard deviation **of the sample**:

Variance, $s^2 = \Sigma d^2/n$ Standard deviation, $s = \sqrt{s^2}$

(c) To estimate the variance and standard deviation of **the population**:

Variance, $\sigma^2 = \Sigma d^2/(n-1)$ Standard deviation, $\sigma = \sqrt{\sigma^2}$

(d) To estimate the standard deviation **of the mean** of the sample:

S.D. of mean, $\sigma_n = \sigma/\sqrt{n}$ (use Table 73 for finding $\sqrt{n}$)

(e) Repeat calculations to check correctness of working.

An example of this type of calculation is shown in Table 54. It

Table 54   **Calculation of variance and standard deviation**

$$\frac{(\Sigma x)^2}{n} = \frac{17 \cdot 67^2}{5} = \frac{312 \cdot 2289}{5} = 62 \cdot 4458$$

x	x^2
3·54	12·5316
3·51	12·3201
3·58	12·8164
3·49	12·1801
3·55	12·6025

$\therefore \Sigma d^2 = 62 \cdot 4507 - 62 \cdot 4458 = 0 \cdot 0049$

$\Sigma x = 17 \cdot 67$ $\Sigma x^2 = 62 \cdot 4507$

$n = 5$

$s^2 = 0 \cdot 0049/5 = 0 \cdot 00098$

$s = \sqrt{0 \cdot 00098} = 0 \cdot 03130$

$\sigma^2 = 0 \cdot 0049/4 = 0 \cdot 001225$

$\sigma = \sqrt{0 \cdot 001225} = 0 \cdot 03500$

$\sigma_n = 0 \cdot 03500/\sqrt{5} = 0 \cdot 01565$

157

illustrates the importance of working in full. Σd^2 is finally found as the very small difference between two larger numbers. If four-figure tables of logarithms or of squares are used for this part of the calculation, the value of $(\Sigma x)^2/n$ may even work out to be greater than Σx^2. Then Σd^2 would be negative, which is absurd. Even if it comes out positive, it will be seriously in error. In this example, calculation of $(\Sigma x)^2/n$ by four-figure tables, gives a value 62·43, making $\Sigma d^2 = 0·02$, which is *over four times* its true value.

2 WHEN THERE ARE MANY OBSERVATIONS

Follow these instructions, referring to the two worked examples in Table 55 and 56:

(a) Set out a table the same as that for calculation of the mean, but with two additional columns, headed fD^2, and $f(D+1)^2$. Fill in the first six columns as in the calculations of the mean (p. 156, paragraphs (a) to (g)).

(b) In the seventh column enter values of fD^2, obtained by multiplying figures in third and fourth columns.

(c) Sum the seventh column, finding ΣfD^2.

(d) In the eighth column, enter $f(D+1)^2$, obtained by multiplying figures in the fifth and sixth columns.

(e) Sum the eighth column, giving $f(D+1)^2$.

(f) As a check on the correctness of working confirm that:

$$\Sigma f(D+1)^2 - \Sigma fD^2 - 2\Sigma fD = \Sigma f$$

If this equation is satisfied, working is correct; if not, examine for errors.

(g) Calculate Σd^2 by the equation:

$$\Sigma d^2 = c^2 \left\{ \Sigma fD^2 - \frac{(\Sigma fD)^2}{\Sigma f} \right\} \quad \text{where} \quad \begin{array}{l} c = \text{class interval} \\ \Sigma f = \text{no. of observations } (=n). \end{array}$$

(h) Use the value of Σd^2 to calculate variances and standard deviations according to the equations given in paragraphs (b) to (d) on p. 157.

(i) The mean may be calculated from: $\bar{x} = w + c\Sigma fD/\Sigma f$, as on p. 156.

(j) Repeat calculations of stages (g) to (i) above to check correctness of working.

Table 55 **Calculation of variance and standard deviation; the number of seeds in ripe pods of Laburnum** (*see* also Table 52)

Observations are not grouped; class interval, $c = 1$; working mean, $w = 3$.

No. of seeds per pod	No. of pods						
x	f	D	fD	$(D+1)$	$f(D+1)$	fD^2	$f(D+1)^2$
1	27	-2	-54	-1	-27	108	27
2	32	-1	-32	0	0	32	0
3	21	0	0	1	21	0	21
4	15	1	15	2	30	15	60
5	13	2	26	3	39	52	117
6	9	3	27	4	36	81	144
7	4	4	16	5	20	64	100
8	1	5	5	6	6	25	36
	$\Sigma f = 122$		$\Sigma fD =$ $89 - 86$ $= 3$			ΣfD^2 $= 377$	$\Sigma f(D+1)^2$ $= 505$

CHECK: $\Sigma f(D+1)^2 - \Sigma fD^2 - 2\Sigma fD = 505 - 377 - (2 \times 3) = 122 = \Sigma f$

Working correct so far.

CALCULATION OF Σd^2:

$$\Sigma d^2 = \Sigma fD^2 - \frac{(\Sigma fD)^2}{\Sigma f} = 377 - \frac{3^2}{122} = 377 - 0.7377 = 376.926$$

$s^2 = 376.926/122 = 3.090$ seeds^2 $\qquad \sigma = \sqrt{3.115} = 1.765$ seeds

$s = \sqrt{3.090} = 1.758$ seeds $\qquad \sigma_n = 1.765/\sqrt{122} = 0.1592$ seeds

$\sigma^2 = 376.926/121 = 3.115$ seeds^2

Table 56 **Calculation of variance and standard deviation; the numbers of flowers on antirrhinum plants** (*see* also Table 53)

Observations grouped; class interval, $c = 20$; working mean is the middle value of the modal class, 60–79, so $w = 69.5$.

No. of flowers per plant	No. of plants						
x	f	D	fD	$(D+1)$	$f(D+1)$	fD^2	$f(D+1)^2$
20– 39	6	-2	-12	-1	-6	24	6
40– 59	10	-1	-10	0	0	10	0
60– 79	12	0	0	1	12	0	12
80– 99	7	1	7	2	14	7	28
100–119	5	2	10	3	15	20	45
120–139	1	3	3	4	4	9	16
140–159	1	4	4	5	5	16	25
160–179	1	5	5	6	6	25	36
	$\Sigma f =$ 43		$\Sigma fD =$ $29 - 22 = 7$			ΣfD^2 $= 111$	$\Sigma f(D+1)^2$ $= 168$

CHECK: $\Sigma f(D+1)^2 - f\Sigma D^2 - 2\Sigma fD = 168 - 111 - (2 \times 7) = 43 = \Sigma f$

Working correct so far.

CALCULATION OF Σd^2:

$$\Sigma d^2 = c^2 \left\{ \Sigma f D^2 - \frac{(\Sigma f D)^2}{\Sigma f} \right\} = 20^2 \left\{ 111 - \frac{7^2}{43} \right\} = 43944{\cdot}0$$

$s^2 = 43944{\cdot}0/43 = 1022$ flowers^2

$s = \sqrt{1022} = 31{\cdot}97$ flowers

$\sigma^2 = 43944{\cdot}0/42 = 1046$ flowers^2

Limits for a mean

(a) Calculate the sample mean ($\bar{x}$) and estimate the standard deviation of sample means (σ_n) according to the methods described in previous sections.

(b) Decide on the level of significance required. For most work, the 5% level is adequate ($p = 0{\cdot}05$), but if greater certainty is needed, choose a higher level of significance (1% or 0·1%) which will give wider limits.

(c) Enter the table of the distribution of t (Table 74, p. 192) at $(n - 1)$ degrees of freedom, where n is the number of observations in the sample. Read the value of t in the column headed by the level of significance decided upon.

(d) Calculate the deviation of limits:

$$\text{deviation of limits} = \pm t\sigma_n$$

(e) Express the best estimate of the population mean in the form:

$$\text{Mean} = \bar{x} \pm t\sigma_n \text{ (level of significance)}$$

Here are two worked examples of this technique:

(*i*) USING THE DATA FROM TABLES 52 AND 55

$$\bar{x} = 3{\cdot}0246 \text{ seeds} \qquad \sigma_n = 0{\cdot}1592 \text{ seeds}$$

There are 122 observations, and 121 degrees of freedom.
From the table, for $p = 0{\cdot}05$, and 120 degrees of freedom (the nearest), $t = 1{\cdot}98$

$$\therefore \text{ deviation of limits} = \pm 1{\cdot}98 \times 0{\cdot}1592 = \pm 0{\cdot}3152$$

Thus the population mean is expressed:

$$\text{Mean} = 3{\cdot}02 \pm 0{\cdot}32 \text{ seeds (5\% significance)}$$

In saying that the mean of the population lies between 2·70 and 3·34 there is only 1 chance in 20 that the statement is wrong.

For $p = 0{\cdot}001$, Table 78 gives $t = 3{\cdot}37$, and the deviation of the limits is $\pm 3{\cdot}37 \times 0{\cdot}1592$, that is $\pm 0{\cdot}5365$.

Thus, one may also express the population mean as:

$$\text{Mean} = 3{\cdot}02 \pm 0{\cdot}54 \text{ seeds (0\cdot1\% significance)}$$

In saying that the mean of the population lies between 2·48 and 3·56 there is only 1 chance in 1000 that the statement is wrong. One can have greater confidence in this statement than in the other one, but the statement itself is less precise, because the limits given are wider. In selecting a level of significance one must generally choose between making a precise statement with a degree of uncertainty, and making a less precise statement which is almost certainly correct.

(*ii*) USING THE DATA FROM TABLES 53 AND 56

$$\bar{x} = 72\cdot756 \text{ flowers} \qquad \sigma_n = \sigma/\sqrt{n} = \frac{\sqrt{1046}}{\sqrt{43}} = 4\cdot933 \text{ flowers}$$

There are 43 observations, and 42 degrees of freedom.
From Table 74, for $p = 0\cdot05$, and 40 degrees of freedom (the nearest), $t = 2\cdot02$.

$$\therefore \text{ deviation of limits} = \pm2\cdot02 \times 4\cdot933 = \pm9\cdot965$$

Thus the population mean is:

$$\text{Mean} = 72\cdot76 \pm9\cdot97 \text{ flowers (5\% significance)}$$

As an alternative to stating the level of significance as a percentage, the value of p may be given, thus:

$$\text{Mean} = 72\cdot76 \pm9\cdot97 \text{ flowers } (p = 0\cdot05)$$

Comparison of two means

(a) Set out two tables for calculating the means and variances of the two samples, according to the instructions in previous sections of this chapter. The mean of the first sample is referred to as $\bar{x}_1$, and that of the second sample as $\bar{x}_2$; similarly the two variances are referred to as σ_1^2 and σ_2^2, and the numbers of observations in each sample are n_1 and n_2.

(b) Calculate the variance of the *difference* of the means:

$$\sigma_d^2 = \frac{\sigma_1^2}{n_1} + \frac{\sigma_2^2}{n_2}$$

(c) Calculate t:

$$t = \frac{x_1 - x_2}{\sigma_d}$$

(If $x_1 < x_2$, transpose them in this equation)

(d) Enter Table 74 at $(n_1 + n_2 - 2)$ degrees of freedom and read the value of t under the chosen level of significance. If the calculated t

exceeds that in the table, the difference between the means is greater than is expected from two random samples drawn from a single population. Therefore the samples are from two populations and their means are significantly different. The confidence placed in this statement is determined by the level of significance chosen when reading the table of *t*.

If the calculated value of *t* is less than that in the table, the two samples may be from a single population, and there is no evidence to the contrary. Their means do not differ significantly.

Refer to pp. 48–49 and pp. 95–9 for a worked example.

Limits for standard deviation

(a) Calculate σ according to the method given on pp. 157–60.

(b) Calculate σ_σ the standard deviation of the standard deviation from:

$$\sigma_\sigma = \frac{\sigma}{\sqrt{(2n)}}, \text{ where } n \text{ is the number of observations.}$$

Table 73 may be used for obtaining $\sqrt{(2n)}$.

(c) Look up *t* (Table 74) for the required level of significance and for $(n-1)$ degrees of freedom.

(d) Calculate the deviation of the limits:

$$\text{Deviation} = \pm t\sigma_\sigma$$

(e) Express the standard deviation in the form:

$$\text{Standard deviation} = \sigma \pm t\sigma_\sigma \text{ (level of significance).}$$

A worked example is given on pp. 52–53.

The variance ratio test, for the comparison of two variances or standard deviations

(a) Calculate σ^2 for each set of observations, as on pp. 157–60.

(b) Calculate the variance ratio: $F = \dfrac{\text{larger variance}}{\text{smaller variance}}$

(c) With n_1 degrees of freedom corresponding to the larger variance and n_2 corresponding to the smaller variance, look up F in one of the Tables 75–78.

(d) If the calculated value of F exceeds that from the table, the variances are significantly different at a level of probability *twice* that given at the head of the table.

For a worked example, see pp. 53–54.

Analysis of variance

There are four main types of analysis of variance described in this book:

1 WITH THREE OR MORE REPLICATED TREATMENTS
(a) Set out the data in a table (see Table 57), sum the columns, and find the grand total.

Table 57 Calculation of sums of squares of data of Table 12: water lost from leaves of Cherry Laurel (mg/cm² in three days)

	Surface covered with jelly					
Replicate No.	Neither	Top	Bottom	Both	Totals	
1	86	41	25	13		
2	108	44	35	11		
3	118	40	37	13		
4	79	52	26	13		
						Grand
Totals	391	177	123	50	741	total
Sums of squares of observations	39225 a_1	7921 a_2	3895 a_3	628 a_4	51669 = A	
Squares of column totals, divided by No. of observations	38220·25 b_1	7832·25 b_2	3782·25 b_3	625·00 b_4	50459·75 = B	
Differences (a − b)	1004·75	88·75	112·75	3·00	1209·25 = A − B	

CHECK: $1004·75 + 88·75 + 112·75 + 3·00 = 1209·25 = 51669 - 50459·75$
Working correct so far.
$D = 741^2/16 = 549081/16 = 34318$

Total SOS $= A - D = 51669 - 34318 = 17351$
Between-
   treatments SOS $= B - D = 50460 - 34318 = 16142$
Residual SOS $= A - B = 51669 - 50460 = 1209$

} Sums of squares required for analysis of variance (see Table 58).

CHECK: $16142 + 1209 = 17351$ Working correct so far.

Statistics for biology

(b) Square each observation, and sum these squares by columns; enter the sums in the spaces marked a_1, a_2, etc. Sum these sums, and call the total A.

(c) Square the column totals, and divide each by the number of observations in each column. Call these b_1, b_2, etc., and enter them in the table. Add them together and call the total B.

(d) Square the grand total, and divide by the total number of observations; call the result D.

(e) Subtract every b from its corresponding a, and enter the differences in the bottom row of the table. Check that the total of the differences, added across the bottom row, is equal to $(A - B)$. This checks the calculation so far, except the grand total and D. Sum the column totals again to check the grand total; repeat stage (d) to check the value of D.

(f) Set out a table of the analysis of variance (Table 58). Check that

Table 58 Analysis of variance of the data from Table 57

Source of variance	Sum of squares	Degrees of freedom	Mean squares
Between-treatments (columns)	$B - D = 16142$	$u - 1 = 3$	5381
Residual	$A - B = 1209$	$u(v - 1) = 12$	101
Total	$A - D = 17351$	$uv - 1 = 15$	

the total sum of squares does in fact agree with the sum of the between-treatments and residual sums of squares. Degrees of freedom are calculated as shown, u being the number of treatments and v being the number of replicates. In this example, $u = v = 4$. The mean square is calculated by dividing the sums of squares by their degrees of freedom.

(g) Apply the variance ratio test:

$$F = \frac{\text{between-treatments mean square}}{\text{residual mean square}}$$

(h) Repeat the calculation of mean squares and variance ratio, as a check.

(i) In the table of variance ratio, find F when n_1 is the number of degrees of freedom of the between-treatments mean square, and n_2 is the number of degrees of freedom of the residual. If the calculated F exceeds this, the variance due to the effects of treatments is significantly greater than the residual variance due to random error.

164

WORKING OF THE EXAMPLE SHOWN IN TABLES 57 AND 58:
(a) $86 + 108 + 118 + 79 = 391$ (entered in Table 57, and so on for every column.
$391 + 177 + 123 + 50 = 741$ (the grand total, entered in Table 57).
(b) $86^2 + 108^2 + 118^2 + 79^2 = 39225$ (entered in space a_1, and so on for every column).
$39225 + 7921 + 3895 + 628 = 51669$ (gives A, entered in Table 57).
(c) $391^2/4 = 38220 \cdot 25$ (entered in space b_1, and so on for every column).
$38220 \cdot 25 + 7832 \cdot 25 + 3782 \cdot 25 + 625 \cdot 0 = 50459 \cdot 75$ (gives B, entered in Table 57).
(d) $741^2/16 = 34318$ (gives D, written below the table).
(e) $39225 - 38220 \cdot 25 = 1004 \cdot 75$ (entered in the bottom row, and so on for every column).
Calculation checked, as shown below the table.
(f) Sums of squares calculated, as shown below Table 57, and transferred to Table 58. Degrees of freedom entered in Table 58.

Between columns mean square $= 16142/3 = 5381$ (entered in Table 58).

Residual mean square $\qquad = 1209/12 = \quad 101$ (entered in Table 58).

(g) $F = 5381/101 = 53$.
(h) Calculation of stages (f) and (g) repeated, and found correct.
(i) In Table 78, when $n_1 = 3$, $n_2 = 12$, and $p = 0 \cdot 001$, $F = 10 \cdot 8$. The calculated F ($= 53$) exceeds this, so the effect of treatment is significant at the $0 \cdot 001 \%$ level.

Making calculation easier: A good way to reduce computational work is to subtract a constant number from every observation when setting out the table for the calculation of sums of squares. This reduces the magnitude of the numbers and makes it easier to calculate squares and find their sums. The figures of Table 59 are derived from those of Table 57 by subtracting 40 from each observation. An analysis of variance performed on these reduced figures will yield the same result, as is demonstrated in the next section.

2 WITH TWO EXPERIMENTAL FACTORS, BUT NO REPLICATION
In this example the data from the previous example have been rearranged according to leaf area, as was done in Table 13. Also, 40 has been subtracted from each observation, before entering it in the table, to make calculation less laborious. The procedure for the analysis is as follows (see worked example in Table 59):

Statistics for biology

Table 59 Calculation of sums of squares of data from Table 13: water lost from leaves of Cherry Laurel (mg/cm² in three days)

All observations have had 40 mg/cm² subtracted from them before being entered in the table.

| Area of leaf | Surface covered with jelly | | | | Totals | Sums of squares of observations |
	Neither	Top	Bottom	Both		
1 (smallest)	46	0	−14	−27	5	3041 a_1'
2	68	4	−15	−27	30	5594 a_2'
3	78	12	−3	−27	60	6966 a_3'
4 (largest)	39	1	−5	−29	6	2388 a_4'
Totals	231	17	−37	−110	101	Grand total
Sums of squares of observations	14345 a_1	161 a_2	455 a_3	3028 a_4		17989 = A

$$B = 16780 \qquad C = 1140 \qquad D = 638$$

$$u = v = 4$$

Total SOS $= A - D$ $= 17989 - 638 = 17351$
Between-columns SOS $= B - D = 16780 - 638$ $= 16142$
Between-rows SOS $= C - D = 1140 - 638$ $=$ 502
Residual SOS $= 17351 - (16142 + 502) =$ 707

(a) Tabulate the data (after subtraction of a constant, if convenient), sum the columns and rows, and find the grand total. Check that the grand total obtained by summing the row totals is the same as that obtained by summing the column totals.

(b) Square the observations, sum the squares by columns, and enter these in the spaces marked a_1, a_2, etc. Sum these sums and call the total A.

(c) Square the observations, sum the squares by rows, and enter these in the spaces marked a_1', a_2', etc. Sum these totals and check that it agrees with the value for A calculated under (b).

(d) Square the column totals, and sum them; divide by the number of observations in each column (v); call the result B.

(e) Square the row totals and sum them; divide by the number of observations in each row (u); call the result C.

(f) Square the grand total; divide by the total number of observations (uv); call the result D.

(g) Repeat stages (d), (e), and (f), to check the working.

(h) Set out a table of the analysis of variance (Table 60). The between-columns, between-rows, and total sum of squares are obtained by

Table 60 Analysis of variance of the data from Table 59

Source of variance	Sum of squares	Degrees of freedom	Mean squares
Between columns (treatments)	$B - D = 16142$	$u - 1 = 3$	5381
Between rows (leaf area)	$C - D = 502$	$v - 1 = 3$	167
Residual	707	$(u-1)(v-1) = 9$	79
Total	$A - D = 17351$	$uv - 1 = 15$	

differences between the quantities A, B, C, and D, as shown beneath Table 59. The residual sum of squares (**SOS**) is obtained from:

Residual SOS = Total SOS − (rows SOS + columns SOS)

Degrees of freedom are calculated as shown in Table 60. Calculate the mean squares.

(i) Apply the variance ratio test to the between-columns mean square and the between-rows mean square:

$$F = \frac{\text{between-columns mean square}}{\text{residual mean square}}$$

and

$$F = \frac{\text{between-rows mean square}}{\text{residual mean square}}$$

(j) Repeat the calculations of stages (h) and (i), as a check.

(k) In the table of variance ratio, find F when n_1 is the number of degrees of freedom of the between-columns mean square, and n_2 is the number of degrees of freedom of the residual. If the F calculated for the between-columns mean square exceeds this, the variance due to differences between columns is significantly greater than the residual variance.

(l) Repeat (k) for the between-rows mean square, to test the significance of variance due to differences between rows.

WORKING OF THE EXAMPLE SHOWN IN TABLES 59 AND 60

(a) $46 + 68 + 78 + 39 = 231$ (entered in Table 59, and so on for every column).

$46 + 0 - 14 - 27 = 5$ (entered in Table 59, and so on for every row).

$231 + 17 - 37 - 110 = 101$ (the grand total, entered in Table 59).

$5 + 30 + 60 + 6 = 101$ (sum of row totals, to check grand total and all calculation so far).

(b) $46^2 + 68^2 + 78^2 + 39^2 = 14345$ (entered in space a_1, and so on for every column.

$14345 + 161 + 455 + 3028 = 17989$ (gives A, entered in Table 59).

(c) $46^2 + 0^2 + (-14)^2 + (-27)^2 = 3041$ (entered in space a_1', and so on for every row).

$3041 + 5594 + 6966 + 2388 = 17989$ (gives A, checks correctness of working so far).

(d) $\dfrac{231^2 + 17^2 + (-37)^2 + (-110)^2}{4} = \dfrac{67119}{4} = 16780$ (gives B, written below table).

(e) $\dfrac{5^2 + 30^2 + 60^2 + 6^2}{4} = \dfrac{4561}{4} = 1140$ (gives C, written below the table).

(f) $101^2/16 = 10201/16 = 638$ (gives D, written below the table).

(g) Calculation of stages (d), (e), and (f) repeated and found correct.

(h) Sums of squares calculated, as shown below Table 59, and transferred to Table 60. Degrees of freedom entered in Table 60.

Between-columns mean square $= 16142/3 = 5381$ (entered in Table 60)

Between-rows mean square $= 502/3 = 167$ (entered in Table 60)

Residual mean square $= 707/9 = 79$ (entered in Table 60)

(i) For between-columns mean square, $F = 5381/79 = 68$.

For between-rows mean square, $F = 167/79 = 2.1$.

(j) Calculation of stages (h) and (i) repeated, and found correct.

(k) In Table 78, when $n_1 = 3$, $n_2 = 9$, and $p = 0.001$, $F = 13.9$. The between-columns variance ratio (68) exceeds this, so variance between columns is significant at the 0.1% level. The covering of different surfaces of the leaf produces highly significant differences in the amounts of water lost.

In Table 76, when $n_1 = 3$, $n_2 = 9$ and $p = 0.05$, $F = 3.9$

and in Table 75, when $n_1 = 3$, $n_2 = 9$ and $p = 0.20$, $F = 1.9$

(l) The between-rows variance ratio (2.1) comes between these values, so variance between rows is significant at the 20% level, but not at the 5% level. This is a slight indication that leaf area has an effect on water loss (greater loss from leaves of medium area), but further experiments are needed to confirm this finding.

Note: Compare the results of this analysis with that of the first example (Tables 57 and 58):

1. Although in this analysis all items of data were reduced by a constant amount (40), the sums of squares and mean squares for between-columns and total variance have come to the same values. This demonstrates that reduction by a constant amount can be used to ease the burden of manipulating large numbers, without upsetting

the analysis in any way. This is an important saving when working without a machine.

2. In both cases the total sum of squares is 17351. In the first example this is partitioned into:

16142—between-columns SOS—due to differences between treatments with jelly.

1209—residual SOS —due to variation within treatments, the random variation from leaf to leaf.

After the calculation of mean squares, a variance ratio of 53 was found, as a measure of the between-column variance with respect to the residual variance. This is significant at the 0·1 % level.

In the second example, the two-factor analysis, the total sum of squares is partitioned into:

16142—between-columns SOS—due to differences between treatments with jelly.

502—between-rows SOS —due to differences between leaves of different area

707 —residual SOS —due to random variation from leaf to leaf

Total of these = 1209

By partitioning the total SOS in this way, part of the residual SOS of the first example has now been shown to be due to difference in leaf area. Though this effect of leaf area does not turn out to be really significant in itself, one has gained by accounting for a part of the total SOS in this way, for one is now left with a smaller sum of squares for the residual. Consequently, the variance ratio for the effect of jelly treatment, which was 53, now becomes 68, and the effect of this treatment is shown to be even more significant than it appeared from the first example.

By allowing for the second variable (leaf area) one has increased the significance of an already highly significant result. This is unimportant in this example, but in many instances it may happen that by allowing for a second factor, and removing its sum of squares from the residual one may so increase the resolving power of the analysis, that the significance of the other factor may be raised from a dubious level to a decidedly significant one. This advantage is additional to the other clear advantage of being able to investigate two factors simultaneously, with a great saving in time, materials, and labour.

The contrast between these two examples emphasises the necessity of designing and performing experiments with as many as possible of the variable factors under control, so that they may be allowed for in the analysis.

3 WITH THREE EXPERIMENTAL FACTORS IN A LATIN SQUARE DESIGN
The basis of the design and a description of the experiment are given on pp. 89–92. The analysis at first follows the same procedure as the two-factor analysis explained above, and is set out in Table 61. A sum of squares for the third factor (designated by different letters in the latin square) is set out as shown in Table 62. The table for the analysis of variance (Table 63) has an additional source of variance, the between-letters variance, which in this example is due to differences between the disc sizes.

Table 61

Calculation of sums of squares: times taken to rise to the surface (s)

Light intensity

Temperature (°C)	1	2	3	4	Totals	Sums of squares of observations
8	17 A	17 C	17 B	24 D	75	1443 a_1'
14	6 B	15 D	4 A	4 C	29	293 a_2'
21	13 D	3 A	5 C	4 B	25	219 a_3'
28	4 C	2 B	5 D	2 A	13	49 a_4'
Totals	40	37	31	34	142	*Grand total*
Sums of squares	510	527	355	612		2004 = A
of observations	a_1	a_2	a_3	a_4		

$B = 1272$ $C = 1815$ $D = 1260$ $E = 1417$ (*see* Table 62)

$$u = v = 4$$

Total SOS = A − D = 2004 − 1260 = 744
Between-columns SOS = B − D = 1272 − 1260 = 12
Between-rows SOS   = C − D = 1815 − 1260 = 555
Between-letters SOS = E − D = 1417 − 1260 = 157
Residual SOS   = 744 − (12 + 555 + 157) = 20

Table 62

Calculation of sums of squares for between-letters variance (data from Table 61)

		Letter		
	A	*B*	*C*	*D*
	17	17	17	24
	4	6	4	15
	3	4	5	13
	2	2	4	5
Totals	26	29	30	57	142 *Grand total*

The procedure in detail:

(a) to (f) As in instructions (a) to (f) on pp. 166–7.

(g) Set out a table of observations under letters, as Table 62. Check agreement of grand total. Square the totals and sum them; divide by the number of observations under each letter (u); call the result E.

(h) Repeat stages (d) to (g), to check working.

(i) Set out a table of the analysis of variance (Table 63). The sums of squares are obtained as in Table 61, the residual sum of squares being obtained by difference.

Table 63

Analysis of variance of the data from Table 61

Source of variance	Sum of squares	Degrees of freedom	Mean squares
Between columns (light)	$B - D = 12$	$u - 1 = 3$	4
Between rows (temperature)	$C - D = 555$	$u - 1 = 3$	185
Between letters (size)	$E - D = 157$	$u - 1 = 3$	51
Residual	20	$(u - 1)(u - 2) = 6$	3·33
Total	$A - D = 744$	$u^2 - 1 = 15$	

(j) Apply the variance ratio test on the mean squares for between-columns, between-rows, *and between letters*, using the method given in instructions (i) to (l) on p. 167. The interpretation of the analysis above is given on pp. 90–1.

4 WITH TWO EXPERIMENTAL FACTORS, AND REPLICATION

The basis of the design and a description of the experiment are given on pp. 93–4. In this analysis the residual SOS is calculated directly, which then makes it possible to find, by difference, an SOS for the interaction between treatments. In general the analysis follows the standard procedure, as shown in Table 64.

171

Table 64

Calculation of sums of squares; times taken to rise to the surface (s)

Light intensity

Temperature ($^\circ C.$)	1	2	3	4	Totals	Sums of squares of cell totals
8	17+9 =**26**	13+15 =**28**	13+16 =**29**	8+10 =**18**	101	2625 α_1'
14	6+8 =**14**	5+6 =**11**	4+6 =**10**	4+5 =**9**	44	498 α_2'
21	3+4 =**7**	3+5 =**8**	3+4 =**7**	5+4 =**9**	31	243 α_3'
28	2+2 =**4**	2+3 =**5**	2+3 =**5**	2+2 =**4**	18	82 α_4'
Totals	51	52	51	40	194	*Grand total*
Sums of squares of cell totals	937 α_1	994 α_2	1015 α_3	502 α_4		3448 = α

$$A = 1724 \qquad B = 1188 \qquad C = 1678 \qquad D = 1176 \qquad L = 1774$$
$$u = v = 4 \qquad \text{no. of replicates} = \rho = 2$$

Total SOS $= L - D = 1744 - 1176 = 598$
Between-columns SOS $= B - D = 1188 - 1176 = 12$
Between-rows SOS $\quad = C - D = 1678 - 1176 = 502$
Residual SOS $\qquad = L - A = 1774 - 1724 = 50$
Interaction SOS $\qquad = 598 - (12 + 502 + 50) = 34$

The procedure is as follows:
(a) Tabulate the data (after subtraction of a constant, if convenient), writing replicate observations together in the cells of the table. Sum each set of replicates and write their totals in the cells of the table. These are referred to as cell totals, and are printed in **bold type** in Table 64.
(b) Sum the cell totals in the columns and rows, and find the grand total. Check that the grand total obtained by summing the row totals is the same as that obtained by summing the column totals.
(b) Square the cell totals, sum the squares by columns, and enter these in the spaces marked α_1, α_2 etc. Sum these sums and call the total α.
(c) Square the cell totals, sum the squares by rows, and enter these in the spaces marked α_1', α_2', etc. Sum these totals and check that it agrees with the value for α calculated under (b).
(d) Divide α by the number of observations in each cell (= the number of replicates, in this example, 2), and call the result A.

(e) Square the column totals, and sum them; divide by the number of *observations* in each column (in this example, 8); call the result *B*.

(f) Square the row totals, and sum them; divide by the number of *observations* in each row; call the result *C*.

(g) Square the grand total, divide by the total number of observations; call the result *D*.

(h) Square each observation, and sum these squares; call the result *L*.

(i) Repeat stages (d) to (h), to check the working.

(j) Set out a table of the analysis of variance (Table 65). The SOS for interaction is obtained by difference, after the other SOSs have been calculated in the manner shown in Table 64. The calculation of degrees of freedom is explained in Table 65.

(k) Apply the variance ratio test, as in section (i) to (l), p. 167. The interpretation of the analysis shown above is given on pp. 93–4.

Table 65 Analysis of variance of the data from Table 64

Source of variance	Sum of squares	Degrees of freedom	Mean squares
Between columns (light)	$B - D = 12$	$u - 1 = 3$	4
Between rows (temperature)	$C - D = 502$	$v - 1 = 3$	167
Interaction (light × temperature)	34	$(u - 1)(v - 1) = 9$	3·8
Residual	$L - A = 50$	$uv(\rho - 1) = 16$	3·1
Total	$L - D = 598$	$uv\rho - 1 = 31$	

Missing values in the analysis of variance

If an observation is missing from the analysis because of accident, gross error, or some other mishap, the missing value may be replaced by the value x, given by:

$$x = \frac{rR + cC - T}{(r - 1)(c - 1)}$$

where r = No. of rows $\qquad$ R = total of row with the missing value

$\quad c$ = No. of columns $\quad$ C = total of column with the missing value

$\qquad\qquad\qquad\qquad T$ = grand total

For example, suppose that one of the leaves of the water-loss experiment had fallen from its support and been trampled on, the data of Table 13 would then have had one missing value, as shown in Table 66. Totalling the rows and columns one obtains the values shown in the table, from which one finds the missing value:

Statistics for biology

Table 66 Data of Table 13 (condensed), with one missing value

				Row totals
86	40	26	13	165
108	x	25	13	146 = R
118	52	37	13	220
79	41	35	11	166
391	133	123	50	697 = T
	= C			

$$r = c = 4$$

$$x = \frac{(4 \times 146) + (4 \times 133) - 697}{3 \times 3} = \frac{419}{9} = 47$$

This value, $x = 47$, can be inserted in the table for the calculation of sums of squares and the analysis completed. However, since the rest of the data has been used in estimating the missing member, the number of degrees of freedom of the residual SOS and the total SOS must both be reduced by 1, when calculating their mean squares and when using the table of F.

If two values are missing, call them x and y. Insert an estimated value for x (one might use the mean of the row or column). Then calculate y, using the equation given above. Insert this value of y, and then calculate a new value for x, using the equation. Insert this value of x, and then calculate a new value for y, and so on until one of the new values of x or y is the same as the previously calculated one. The last two values of x and y may then be inserted and the analysis of variance continued, with the loss of two degrees of freedom from each of the sum of squares of the residual and the total.

For example, if two observations are missing from the data of Table 13, one might get Table 67. Estimate the value of x, say, 45.

Table 67 Data of Table 13 (condensed), with two missing values

				Row totals
86	40	26	13	165
108	x	25	13	146
118	52	y	13	183
79	41	35	11	166
391	133	86	50	660 = T

$$r = c = 4$$

$r = c = 4$ throughout the calculation. Then for the row and column containing y,

$R = 183$, $C = 86$, $T = 660 + 45 = 705$. This gives $y = 41$

Insert $y = 41$ in the table. Now for the row and column containing x,

$R = 146$, $C = 133$, $T = 660 + 41 = 701$. This gives $x = 46$

Insert $x = 46$ in the table. Now for the row and column containing y,

$R = 183$, $C = 86$, $T = 660 + 46 = 706$. This gives $y = 41$

This is the same as the previous value of y, so the calculation is complete, and the missing values to be inserted are: $x = 46$, and $y = 41$.

Coefficient of correlation

(a) Set out the data in a table, as in Table 68.

Table 68 Calculation of coefficient of correlation of data from Table 14

Weight (g)	Length (cm)			
x	y	x^2	y^2	xy
0·7	1·7	0·49	2·89	1·19
1·2	2·2	1·44	4·84	2·64
0·9	2·0	0·81	4·00	1·80
1·4	2·3	1·96	5·29	3·22
1·2	2·4	1·44	5·76	2·88
1·1	2·2	1·21	4·84	2·42
1·0	2·0	1·00	4·00	2·00
0·9	1·9	0·81	3·61	1·71
1·0	2·1	1·00	4·41	2·10
0·8	1·6	0·64	2·56	1·28
$\Sigma x = 10\cdot2$	$\Sigma y = 20\cdot4$	$\Sigma x^2 = 10\cdot8$	$\Sigma y^2 = 42\cdot20$	$\Sigma xy = 21\cdot24$

$$n = 10$$

$\Sigma d_a^2 = 10\cdot80 - 10\cdot2^2/10 = 0\cdot396$

$\Sigma d_v^2 = 42\cdot20 - 20\cdot4^2/10 = 0\cdot584$

$\Sigma d_x d_y = 21\cdot24 - (10\cdot2 \times 20\cdot4)/10 = 0\cdot432$

$r \quad = \dfrac{0\cdot432}{\sqrt{(0\cdot396 \times 0\cdot584)}} = \dfrac{0\cdot432}{\sqrt{0\cdot2313}} = 0\cdot8983$

(b) Calculate x^2, y^2, and xy. Sum the columns. Repeat to check working.

(c) Calculate $\Sigma d_x^2 = \Sigma x^2 - (\Sigma x)^2/n$.

(d) Calculate $\Sigma d_v^2 = \Sigma y^2 - (\Sigma y)^2/n$.

(e) Calculate $\Sigma d_x d_y = \Sigma xy - \Sigma x \cdot \Sigma y/n$.

$\Sigma d_x d_y$ can be either negative or positive, depending upon whether the correlation is negative or positive.

(f) Calculate $r = \dfrac{\Sigma d_x d_y}{\sqrt{(\Sigma d^2{}_x \cdot \Sigma d^2{}_y)}}$.

(g) Check the calculations of stages (c), to (f).

(h) Look up r in Table 79, for $(n - 1)$ degrees of freedom, and at the desired level of probability. If the calculated r is negative, its sign is ignored when using the table.

In the example, $(n - 1) = 9$ and when $p = 0.01$, $r = 0.735$. The calculated value of r exceeds this, so there is positive correlation between the variables, significant at the 1 % level. Increased weight is correlated with increased length

Regression equations

FOR THE REGRESSION OF y ON x:

(a) Calculate Σd_x^2, Σd_y^2, and $\Sigma d_x d_y$, as above.

(b) Calculate $\bar{x} = \Sigma x/n = 10.2/10 = 1.02$

(c) Calculate $\bar{y} = \Sigma y/n = 20.4/10 = 2.04$

(d) Calculate $b = \Sigma d_x d_y/\Sigma d_x^2 = 0.432/0.396 = 1.091$

(e) The equation for the regression of y on x is

$$
\begin{aligned}
y &= \bar{y} + b(x - \bar{x}) \\
&= 2.04 + 1.091(x - 1.02) \\
&= 2.04 + 1.091x - 1.113 \\
\therefore \quad y &= 1.091x + 0.927
\end{aligned}
$$

This line has been plotted in Fig. 11A, and an example of the use of this equation is given on p. 69, where a value of y is estimated from a given value of x.

FOR THE REGRESSION OF x ON y:

Work stages (a), (b), and (c), as above.

(d) Calculate $b = \Sigma d_x d_y/\Sigma d_y^2 = 0.432/0.584 = 0.7397$.

(e) The equation for the regression of x on y is

$$
\begin{aligned}
x &= \bar{x} + b(y - \bar{y}) \\
&= 1.02 + 0.7397(y - 2.04) \\
&= 1.02 + 0.7397y - 1.509 \\
\therefore \quad x &= 0.7397y - 0.4890
\end{aligned}
$$

This line has been plotted in Fig. 11B, and an example of its use for estimating a value of x from a given value of y is shown on p. 69.

The χ^2 test

WITH A $1 \times n$ CONTINGENCY TABLE:
(a) Set out the results in a table, as on pp. 74–75 (Tables 15 and 16). If any of the categories contain less than five individuals, they must be amalgamated with one of the other categories. For example, if trap B had caught only four mice it is necessary to amalgamate the figure with that of another trap, say one which was close by it. If this was trap A, the table would then have four categories: Trap A and B (23 + 4 = 27 mice), Trap C (25 mice), Trap D (19 mice), Trap E (21 mice). Unfortunately it would not now be possible to examine the data to see if B had caught significantly fewer mice. If this information was required, the only solution would be to trap for a longer period, so increasing the numbers in all traps, until the least number caught in any trap was five or more.
(b) Calculate the total, and divide this total into categories in the expected ratio, usually the ratio expected in a genetical experiment, or equal numbers in every category, as in Table 16.
(c) Calculate $(O - E)$, $(O - E)^2$, and $(O - E)^2/E$, for every category.
(d) Sum the values of $(O - E)^2/E$ to find χ^2.
(e) Check all calculations so far, by repetition.
(f) Calculate the number of degrees of freedom, which is one less than the number of categories.
(g) In Table 80, look along the row corresponding to the number of degrees of freedom, and pick out those two adjacent columns which show values of χ^2 slightly greater and slightly less than the value of χ^2 obtained by calculation. The values of p given at the heads of these columns gives the probabilities that such a deviation as that observed could have arisen by chance. The two probabilities taken from the table give a measure of the probability of the calculated χ^2, and hence of the observed deviations. If the calculated value of χ^2 exceeds that in the column headed $p = 0.001$, the deviation is extremely significant.

When testing genetical ratios one is generally looking for *good* agreement between observation and expectation, that is, for a high value of p. For a significant result, one hopes to obtain a value for χ^2 close to those given in the left columns of the table, where $p = 0.99$, 0.98, or 0.95.

WITH A 2×2 CONTINGENCY TABLE:
(a) Set out the results in a table, as on p. 76 (Table 17). If any of the entries in the table are less than five, no analysis can be done. Collect

more data. if this is possible, until every entry is greater than five.
(b) The entries in the tables are designated by the symbols given in
Table 69. Calculate χ^2 by the short-cut equation:

$$\chi^2 = \frac{n(ad - bc - n/2)^2}{(a+b)(c+d)(a+c)(b+d)}$$

Only the first three or four digits are needed in the answer. The
item $n/2$ is a correction to allow for the fact that the variable is discontinuous, whereas the test assumes a continuous variable. This
correction may be omitted if there are several hundreds in the samples.

Table 69 **2 × 2 contingency table**

			Totals
	a	*b*	*a+b*
	c	*d*	*c+d*
Totals	*a+c*	*b+d*	*a+b+c+d=n*

In the example of Table 17, $a = 64$, $b = 16$, $c = 34$, $d = 46$, and
$n = 160$

$$\therefore \ \chi^2 = \frac{160(64 \times 46 - 16 \times 34 - 160/2)^2}{80 \times 80 \times 98 \times 62} = \frac{2320^2}{243040} = 22 \cdot 1$$

This is slightly less than the value calculated from first principles
on p. 77, since for the sake of simplifying the explanation, the
correction for continuity was left out.
(c) Check the calculations of stage (b).
(d) In Table 80, with one degree of freedom, find the two adjacent
columns which show values of χ^2 slightly greater and slightly less
than the calculated value. The values of p at the heads of these two
columns give the probability that the two factors are independent
of one another, or show no association (that is, the results are in
agreement with the null hypothesis of independence). Generally one
is looking for *dis*agreement with the null hypothesis, to show that the
two factors are dependent upon one another (for example, that
viable seeds are mainly from Batch A, not from both batches
equally). One hopes to obtain a value for χ^2 close to those in the right
columns of the table, where $p = 0 \cdot 05$, $0 \cdot 02$, $0 \cdot 01$, or $0 \cdot 001$.

Distribution-free tests

Since the procedure for these tests is relatively simple, the reader is
referred to the descriptions given in Chapter 11.

14 Using a calculating machine

In the instructions which follow, it is assumed that the reader knows how to add, subtract, multiply, and divide on a machine. Different makes of machine vary in their construction and capabilities but these instructions are applicable to most types of hand-powered machine. Users of electric machines, with special automatic facilities, will be able to adapt the instructions to their machines.

The following terms and abbreviations are used:

Setting register (SR): the keyboard or system of levers used for entering a number on the machine. Most machines have a control register or indicator to show what number is in the setting register.

Product register (PR): the register which gives the answer to addition, subtraction, and multiplication operations. Sometimes called the results register.

Multiplier register (MR): this indicates how many operations have been performed (the number of turns of the handle). After multiplication, the multiplier appears in MR; in division, the quotient appears there. It is assumed, in these instructions, that MR has tens transmission, that is, tens carry over from one digit to the next on the left. Sometimes called the counter register, or quotient register.

The instructions given in this chapter are specially written for a small hand calculator, like the Multo Models 113 and 115, which have 10 figures in SR, 13 figures in PR, and 8 figures in MR. All the calculations in this book were performed on such a machine. The chapter gives only *special* techniques for use with a machine; if no special methods are available, follow the instructions for calculation in Chapter 12; with a machine, they will be less burdensome.

Means

(a) Set the carriage to the right, in the usual position for division. Clear all registers.

(b) Enter the observations one at a time on SR, adding them together by turning the handle once each time (or pressing the 'add' button

179

on an electric machine). The observations should be entered as far to the left of SR as will allow sufficient room in PR for carrying over of figures. At the end of this operation, PR will contain Σx, and MR will contain Σf ($=n$), showing that n items have been added.

(c) Clear all registers except PR. Enter n in SR, in a position suitable for division. After dividing, the mean $\bar{x}$ will appear in MR.

If negative values are to be included in Σx, they may be subtracted from the cumulative total in PR by turning the handle backwards (or pressing the 'subtract' button on an electric machine). This *reduces* the total of items in MR by one, so that a proper count of items will not be kept, and Σf will be wrong. To keep count of items after every negative number has been subtracted, clear SR, turn the handle forward *twice*. This adds two to the total in MR, and so restores the count to its proper value.

Alternatively, negative values may be dealt with by summing them separately, and then subtracting their total from the total of the positive values. This is the quicker method if there are many negative values.

Grouped data can be treated according to (a), (b), and (c) above, turning the handle once for every observation within the group (that is, f times). For example if the observation '28' has a frequency 4, enter 28 in SR and turn the handle four times. This will add 112 ($=fx$) to the cumulative total in PR, and will add four to the cumulative total in MR. When many frequencies are greater than ten, this procedure becomes tedious, and there is a risk of mis-counting the number of turns. A better technique follows; using as an example the data of Table 55:

(a) Set 1 on the extreme right of SR, and x on the left, allowing room for the full number of digits expected. In this example, x never has more than one digit so the extreme left of SR may be used. The first value of x in the example is 1, so the setting of SR will be:

$$1\ 000\ 000\ 001 \qquad \text{(if the SR has 10 figures)}$$

(b) With the carriage to the left, as for multiplication, multiply the entry in SR by f, in this case by 27. PR will then show:

$$0\ 027\ 000\ 000\ 027 \qquad \text{(if the PR has 13 figures)}$$

MR will show the multiplier, 27.

(c) Clear MR **(but not PR)** and return the carriage to the left. Set the next value of x in place of the old one, leaving the '1' on the right of SR. SR will then show:

$$2\ 000\ 000\ 001$$

(d) Multiply by f, this time 32, after which PR will show:

$$0\ 091\ 000\ 000\ 059$$

MR will show the multiplier, 32. Clear MR, **(but not PR)** and repeat the sequence of stages (c) and (d) for succeeding values of x and f in turn. When the last pair of values has been multiplied, PR will read:

$$0\ 369\ 000\ 000\ 122$$

This is interpreted as $\Sigma fx = 369$, and $\Sigma f = 122$. The two quantities have been summed simultaneously.

(e) Record Σf, return the carriage to the left and enter Σf on the extreme right of SR, subtract it once leaving:

$$0\ 369\ 000\ 000\ 000$$

This leaves only Σfx in PR, ready for calculation of the mean.

(f) Clear SR and MR, put the carriage to the right, enter Σf in SR and divide. The mean appears in MR, which shows:

$$03\ 024\ 590 \qquad \text{(if the MR has 8 figures)}$$

Inspection will indicate the position of the decimal point, which in this case comes between the sixth and seventh figures from the right The value of x is $3 \cdot 024590$.

The explanation of this procedure is that in SR one has placed the value, $(x \cdot 10^9 + 1)$. This has been multiplied by f, giving in PR, $(fx \cdot 10^9 + f)$. These products have been left in PR after every multiplication, so one has accumulated the totals and at the end has $(\Sigma fx \cdot 10^9 + \Sigma f)$. Since in this expression Σfx is multiplied by the large factor 10^9 ($= 1\,000\,000\,000$), it is easily distinguishable at the left of PR from Σf on the right.

Variance and standard deviation

The technique is similar in principle to the last one in the previous section. As an example, take the data of Table 54.

(a) Set 1 on the extreme left of SR, and x on the extreme right. The setting represents $(10^q + x)$ where q is a constant depending upon the capacity of SR and the position of the decimal point. In this example, the first setting is:

$$1\ 000\ 000\ 354$$

Here $q = 7$, since the number on the left is $10\,000\,000 \cdot 00$, or 10^7.

(b) Multiply by x, that is by 3·54. After multiplication there is in MR and PR:

$$0 \ 000 \ 000 \ 354 \qquad \text{and} \qquad 0 \ 354 \ 000 \ 125 \ 316$$
$$\text{MR} \qquad\qquad\qquad\qquad \text{PR}$$

The figures on the left of PR (0 354) give the value of x, and those on the right (000 125 316) give the value of x^2 (actually 12·5316, but it is easier to put in the decimal places at the end).

(c) Clear SR and MR, but not PR; set 1 on the left, and the next observation on the right, as before. In this example, the setting would be 1 000 000 351. Multiply by x (351). The product will be added to the total already present in PR, giving:

$$0 \ 705 \ 000 \ 248 \ 517$$

The 0 705 indicates that the total of x, so far, is 7·05, and the 000 248 517 indicates that the total sum of squares, so far, is 24·8157.

(d) Repeat (c) for each observation in turn. In the example, the final reading will be:

$$1 \ 767 \ 000 \ 624 \ 507$$

This is $\Sigma x \, . \, 10^7 + \Sigma x^2$; so $\Sigma x = 17 \cdot 67$
$$\text{and } \Sigma x^2 = 62 \cdot 4507$$

(e) Continue the calculation of Σd^2, and of the variance and standard deviation using the equations given on p. 157, paragraphs (a) to (d). Square roots may be taken from tables when required; alternatively they may be found on the machine, according to the following method.

CALCULATION OF SQUARE ROOTS BY MACHINE

(a) The machine is operated as for normal division. The number of which the square root is to be found is divided by an estimate of the square root (which may be taken from a table, if it is desired to calculate square roots with greater accuracy than those given in the table.)

For example, to find the square root of 26·932, to four decimal places, first estimate the square root as 5·1. 26·932 (in PR) divided by 5·1 (in SR) gives 5·28 (in MR).

(b) Estimate the mean of the divisor and quotient. In this example, the mean of 5·1 and 5·28 is approximately 5·19—there is no need to bother with an exact figure.

(c) Reset the machine for the next division. This can be done by clearing all registers, but a quicker way is to leave registers as they were at the end of stage (a), return the carriage to the left one position at a time, and in each position turning the handle of the machine forwards until zero appears in MR. This is really carrying out the division operation backwards; at the end, MR will show zero, and the original figure (26·932) will reappear in PR.

(d) Divide next by the new estimate of square root (5·19). The answer is 5·1892. The mean of this and 5·19, is 5·1896.

(e) Reset the machine again, and divide by the new estimate (5·1896). This gives 5·18960. To four decimal places, this has the same value as the divisor, so the square root is 5·1896.

(f) Check by squaring this square root.

This procedure of successive approximation may be continued farther if more decimal places are required, limited only by the capacity of the machine.

VARIANCE AND STANDARD DEVIATION OF GROUPED DATA
The method given on p. 158 requires a multiplication operation for every observation. Even with the comparatively small sample of Table 55, there would be 122 multiplications, and a quicker technique is to be preferred.

(a) With the carriage on the left, set x on the extreme left of SR, and x^2 on the extreme right. Values of x^2 are taken from Table 72. If the data has more than three figures, rounding off to three figures will not usually cause serious loss.

(b) Multiply by the frequency, f.

(c) Clear MR and SR, but not PR; set the next pair of values of x and x^2; multiply by the corresponding f. Do this for every class in turn. In the end PR will indicate Σfx and Σfx^2.

For example, using the data from Table 55, the first setting of SR would be:

$$1\ 000\ 000\ 001 \qquad (x = 1, \text{ and } x^2 = 1)$$

multiplied by $f\ (=27)$ the product in PR would be:

$$0\ 027\ 000\ 000\ 027$$

The next setting of SR would be:

$$2\ 000\ 000\ 004 \qquad (x = 2, \text{ and } x^2 = 4)$$

Multiplied by $f\ (=32)$ and added to the figures already in PR, this would give:

$$0\ 091\ 000\ 000\ 155 \qquad (\text{so far, } \Sigma fx = 91, \text{ and } \Sigma fx^2 = 155)$$

The final total in PR will be:

$$0\ 369\ 000\ 001\ 493 \qquad (\Sigma fx = 369, \text{ and } \Sigma fx^2 = 1493)$$

183

$n = \Sigma fx$ must be calculated separately. From these values calculate Σd^2, and hence the variances and standard deviations, using the equations in paragraphs (a) to (d), p. 157.

Checking calculations

The simplest method is to repeat the calculation. A machine is so reliable and rapid in operation that repetition is not so time-consuming as it is with calculations on paper. Sometimes it is possible to cross-check—for instance, Σx obtained from the calculation of a mean can be checked against Σx obtained from calculation of variance. This is a complete check on the value of Σx and Σf used for the mean, provided that the '1' has always been properly set on the right of SR for the whole calculation. It is not a check on Σx^2, but if care is taken to compare the multiplier (in MR) with the multiplicand (in SR, or shown in the control register) at the completion of every multiplication and it is ascertained that they are identical, there is no need for a further check on Σx^2.

Analysis of variance

The general procedure follows that already described in Chapter 10. Some of the symbols used (G, J and K) are new; they are introduced because the convenient 'stopping-off points' in the calculation by machine are slightly different from those in a manual calculation. G is the grand total, H is the same as A, J/u is the same as B, K/v is the same as C, and G^2/uv is the same as D.

(a) Set out the data in a table (*see* Table 70).

(b) Set the carriage to the left, as for normal multiplication. Enter 1 in the extreme left of SR, and the item in the first row of the first column on the extreme right; multiply by the same item. This gives Σx on the left of PR, and Σx^2 on the right, as explained on p. 182. In this example, PR would read:

0 086 000 007396, indicating that, so far, $\Sigma x = 86$, and $\Sigma x^2 = 7396$.

(c) Clear SR and MR, but not PR, return the carriage to the left, and using the item in the second row of the first column, repeat stage (b). Proceed in this manner down the first column. After reaching the bottom of the first column, PR would read:

0 391 000 039 225 (for the first column, $\Sigma x = 391$, and $\Sigma x^2 = 39\,225$).

Enter these values in the table, and clear the machine.

(d) Repeat stages (b) and (c) for the other columns in turn.

(e) In the same way work along each row of the table in turn, entering Σx and Σx^2 for every row. Clear the machine.

(f) Enter 1 in the extreme left of SR, and the total (Σx) of the 1st column on the extreme right of SR. Multiply by the total. Continue in this way working along the row of column totals. At the end, PR will contain the grand total, G, on the left, and the sum of squares of column totals on the right. Record the grand total in the space marked G, and write the sum of squares of column totals below the table, calling this J. Clear the machine.

(g) In the same way as in (f) work down the column of row totals to obtain the grand total, G, and the sum of squares of row totals. Check that the grand total agrees with that calculated above; if it does, all row and column totals and the grand total are correct; if it does not, check for error. Record the sum of squares of row totals beneath the table, calling this K. Clear the machine.

(h) Work along the sums of squares *row*, by simple addition, to obtain the total sum of squares of all observations; enter this in space H. As a check, work down the sums of squares *column*; this should give the same total, H. If it does, all calculation is correct so far; if not, check for error.

(i) Divide J by v, the number of rows; divide K by u, the number of columns; calculate the square of G, and divide it by uv, the total number of observations. Since the divisions are by small integers, it is often more convenient to *multiply* by the *reciprocal*, for on a hand machine multiplication is quicker than division. Instead of dividing by four, multiply by 0·25, and so on.

At this stage, the calculation has reached that shown in Table 70.

Table 70

Machine calculation of sums of squares of data from Table 13: water lost from leaves of Cherry Laurel (mg/cm² in three days)

| Area of leaf | Surface covered with jelly | | | | | Sums of squares |
	Neither	Top	Bottom	Both	Totals	of observations
1 (smallest)	86	40	26	13	165	9841
2	108	44	25	13	190	14394
3	118	52	37	13	220	18166
4 (largest)	79	41	35	11	166	9268
Totals	391	177	123	50	741 = G	

Table 70 (*continued*)

Sums of
squares of   39 225   7921   3 895   628   51 669 = H
observations

$$J = 201 839 \qquad\qquad K = 139 281$$
$$u = v = 4$$
$$J/4 = \;\; 50 460 \qquad\qquad K/4 = \;\; 34 820$$
$$G^2/16 = 34 318$$

Total SOS   = H $- G^2/16 = 51 669 - 34 318 = 17 351$
Between-columns SOS = $J/4$ $- G^2/16 = 50 460 - 34 318 = 16 142$
Between-rows SOS   = $K/4 - G^2/16 = 34 820 - 34 318 = \;\; 502$
Residual SOS $- 17 351 - (16 142 + 502)$ =   707

Table 71
Analysis of variance of the data from Table 55 (Machine calculation)

Source of variance	Sum of squares	Degrees of freedom	Mean squares
Between-columns (treatments)	$J/v - G^2/uv = 16 142$	$u-1 = 3$	5 381
Between-rows (leaf area)	$K/u - G^2/uv = \;\; 502$	$v-1 = 3$	167
Residual	707	$(u-1)(v-1) = 9$	79
Total	$H - G^2/uv = 17 351$		

(j) Set out a table for the analysis of variance (Table 71). To calculate the sums of squares, enter G^2/uv in SR, turn the handle *backwards*, subtracting G^2/uv from zero once. This gives the complement in PR: 9 999 999 965 682. Put J/v in SR, and *add* this to the complement in PR. This will give the required difference, in this case 16 142.

Without clearing SR or PR, turn the handle backwards to subtract J/v, and so restore the complement of G^2/uv in PR. Put K/u in SR, add and thus find the required difference, $K/u - G^2/uv$. Restore the complement again, add H, and thus find $H - G^2/uv$. Leave this in PR, and from it subtract the between-columns SOS, and the between-rows SOS, leaving the residual SOS in PR. Calculate mean squares.

Table 71, though obtained by a different routine of calculation, gives exactly the same values as those in Table 60. The method of assessing the significance of the variances is now the same as on p. 143, stages (k) and (l). Before doing this, check all calculations of stages (i) and (j) of this calculation.

Notes: 1 The technique of subtracting a constant from all observations, as was done in Table 59, is not worth while when using a machine if it produces negative values, since they cause so much complication in the calculation. In other instances it may be helpful, for instance if the observations are 11234, 11245, 11256, 11313, and so on, they should be reduced to 34, 45, 56, and 113, by the subtraction of 11200 from each. This saves time and machine-space.

2 With a single-factor analysis, with replication (as in Table 57), it is still worth while to calculate *row* totals and sums of squares, as these give a check on the correctness of calculations up to stage (h). Calculations involving K are of course omitted, and the setting out of the analysis of variance and the assessment of significance follows the pattern of the example on p. 164.

Coefficient of correlation, and regression equations

One needs the values of Σx, Σy, Σxy, Σx^2, and Σy^2. There are several ways of calculating these, and the following is the best for a machine of small capacity. The data of Table 14 will be taken as an example.
(a) Enter 1 on the extreme left of SR, x in the centre, and y on the extreme right. The best position for x depends on the capacity of the machine. For a SR with ten-figure capacity, set 1 000 070 017, for the first pair of figures in Table 14. Considering the decimal point to be between the first and second figures on the right, this is equivalent to $(10^8 + x.10^4 + y)$.
(b) Multiply by x. This will give 0 007 000 490 119 in PR, which is equivalent to $x.(10^8 + x.10^4 + y) = x.10^8 + x^2.10^4 + xy$.
Reading from PR:

$$0\ 007|000\ 49\ \ |0\ 119$$
$$x = 0{\cdot}7|x^2 = 0{\cdot}49|xy = 1{\cdot}19$$

(c) Clear MR, and SR, **but not PR**. Repeat (b) for all other pairs of observations. The totals will accumulate in PR. At the end, PR will contain $(\Sigma x.10^8 + \Sigma x^2.10^4 + \Sigma xy)$.
In this example, PR will read:

$$0\ 102\ 010\ 802\ 124$$

Giving:

$$\Sigma x = 10{\cdot}2,\ \Sigma x^2 = 10{\cdot}80,\ \text{and}\ \Sigma xy = 21{\cdot}24$$

(d) Carry out a similar procedure to that in stages (a) to (c), but this time, having entered 1 on the extreme left, x in the centre, and y to the extreme right, multiply by y. This will eventually give Σy, Σxy, and Σy^2, in order across PR. The value of Σxy has been calculated again this time, and gives a check on the correctness of both sets of calculations. If the values of Σxy do not agree, one or both sets of calculations are in error, and must be examined.

(e) Using the values of Σx, Σy, Σxy, Σx^2, and Σy^2 calculated above, calculate r and the regression equations, using the methods given on pp. 175–6.

When using this technique, watch the cumulative totals in PR to make sure that figures from one total are not carried over to the total on its left. If this seems likely to happen, sum the data in two or more parts, record the partial totals and add them at the end.

Unfortunately, if either of the variables has more than two digits, this method becomes limited by the capacity of the machine. If a machine with larger capacity is not available, it is necessary to calculate Σx and Σy by summing both simultaneously, one on the left and the other on the right of SR. Check this calculation by repetition, then set x on the left, y on the right, and multiply by x. Repeat for all pairs of observation, without clearing PR. This gives Σxy and Σx^2. Finally set x on the left, y on the right, and multiply by y. Repeat for all pairs of observations, without clearing PR. This gives Σxy and Σy^2. The comparison of Σxy with the previously calculated value checks the correctness of Σxy, Σx^2, and Σy^2. A more cumbersome technique, but better than working each quantity separately.

Fisher statistical slide rule

This puts the tables for t, F, r, χ^2, and a few other quantities in a handy and easily understood form. Though less precise and more restricted in scope than Tables 74–80, the rule is a useful aid to assessing the significance of results. It is published by the University of London Press, complete with instruction booklet. When using it with the tests described in *Statistics for Biology*:

(a) Use the 'one-tail' windows for F (in the analysis of variance) and for χ^2.

(b) Use the 'two-tail' windows for t, F (when comparing two variances, pp. 52–53) and for r.

(c) Interpret the colours like this:

Red $\quad p < 0.01 \quad$ A significant effect.

Yellow $0.05 > p > 0.01$ Significant or not, depending on the level you have adopted for significance.

Green $\quad p > 0.05 \quad$ Effect not significant. When testing genetical ratios, this indicates that the observed ratio does not differ significantly from that expected.

tatistical Tables

quares of integers, 1–999

	0	1	2	3	4	5	6	7	8	9
0	0	1	4	9	16	25	36	49	64	81
0	100	121	144	169	196	225	256	289	324	361
0	400	441	484	529	576	625	676	729	784	841
0	900	961	1024	1089	1156	1225	1296	1369	1444	1521
0	1600	1681	1764	1849	1936	2025	2116	2209	2304	2401
0	2500	2601	2704	2809	2916	3025	3136	3249	3364	3481
0	3600	3721	3844	3969	4096	4225	4356	4489	4624	4761
0	4900	5041	5184	5329	5476	5625	5776	5929	6084	6241
0	6400	6561	6724	6889	7056	7225	7396	7569	7744	7921
0	8100	8281	8464	8649	8836	9025	9216	9409	9604	9801
0	10000	10201	10404	10609	10816	11025	11236	11449	11664	11881
0	12100	12321	12544	12769	12996	13225	13456	13689	13924	14161
0	14400	14641	14884	15129	15376	15625	15876	16129	16384	16641
0	16900	17161	17424	17689	17956	18225	18496	18769	19044	19321
0	19600	19881	20164	20449	20736	21025	21316	21609	21904	22201
0	22500	22801	23104	23409	23716	24025	24336	24649	24964	25281
0	25600	25921	26244	26569	26896	27225	27556	27889	28224	28561
0	28900	29241	29584	29929	30276	30625	30976	31329	31684	32041
0	32400	32761	33124	33489	33856	34225	34596	34969	35344	35721
0	36100	36481	36864	37249	37636	38025	38416	38809	39204	39601
0	40000	40401	40804	41209	41616	42025	42436	42849	43264	43681
0	44100	44521	44944	45369	45796	46225	46656	47089	47524	47961
0	48400	48841	49284	49729	50176	50625	51076	51529	51984	52441
0	52900	53361	53824	54289	54756	55225	55696	56169	56644	57121
0	57600	58081	58564	59049	59536	60025	60516	61009	61504	62001
0	62500	63001	63504	64009	64516	65025	65536	66049	66564	67081
0	67600	68121	68644	69169	69696	70225	70756	71289	71824	72361
0	72900	73441	73984	74529	75076	75625	76176	76729	77284	77841
0	78400	78961	79524	80089	80656	81225	81796	82369	82944	83521
0	84100	84681	85264	85849	86436	87025	87616	88209	88804	89401
0	90000	90601	91204	91809	92416	93025	93636	94249	94864	95481
0	96100	96721	97344	97969	98596	99225	99856	100489	101124	101761
0	102400	103041	103684	104329	104976	105625	106276	106929	107584	108241

Table 72—*continued*

Squares of integers, 1–999

	0	1	2	3	4	5	6	7	8	9
330	108900	109561	110224	110889	111556	112225	112896	113569	114244	11492
340	115600	116281	116964	117649	118336	119025	119716	120409	121104	12180
350	122500	123201	123904	124609	125316	126025	126736	127449	128164	12888
360	129600	130321	131044	131769	132496	133225	133956	134689	135424	13616
370	136900	137641	138384	139129	139876	140625	141376	142129	142884	14364
380	144400	145161	145924	146689	147456	148225	148996	149769	150544	15132
390	152100	152881	153664	154449	155236	156025	156816	157609	158404	15920
400	160000	160801	161604	162409	163216	164025	164836	165649	166464	16728
410	168100	168921	169744	170569	171396	172225	173056	173889	174724	17556
420	176400	177241	178084	178929	179776	180625	181476	182329	183184	18404
430	184900	185761	186624	187489	188356	189225	190096	190969	191844	19272
440	193600	194481	195364	196249	197136	198025	198916	199809	200704	20160
450	202500	203401	204304	205209	206116	207025	207936	208849	209764	21068
460	211600	212521	213444	214369	215296	216225	217156	218089	219024	21996
470	220900	221841	222784	223729	224676	225625	226576	227529	228484	22944
480	230400	231361	232324	233289	234256	235225	236196	237169	238144	23912
490	240100	241081	242064	243049	244036	245025	246016	247009	248004	24900
500	250000	251001	252004	253009	254016	255025	256036	257049	258064	25908
510	260100	261121	262144	263169	264196	265225	266256	267289	268324	26936
520	270400	271441	272484	273529	274576	275625	276676	277729	278784	27984
530	280900	281961	283024	284089	285156	286225	287296	288369	289444	29052
540	291600	292681	293764	294849	295936	297025	298116	299209	300304	30140
550	302500	303601	304704	305809	306916	308025	309136	310249	311364	31248
560	313600	314721	315844	316969	318096	319225	320356	321489	322624	32376
570	324900	326041	327184	328329	329476	330625	331776	332929	334084	33524
580	336400	337561	338724	339889	341056	342225	343396	344569	345744	34692
590	348100	349281	350464	351649	352836	354025	355216	356409	357604	35880
600	360000	361201	362404	363609	364816	366025	367236	368449	369664	37088
610	372100	373321	374544	375769	376996	378225	379456	380689	381924	38316
620	384400	385641	386884	388129	389376	390625	391876	393129	394384	39564
630	396900	398161	399424	400689	401956	403225	404496	405769	407044	40832
640	409600	410881	412164	413449	414736	416025	417316	418609	419904	42120
650	422500	423801	425104	426409	427716	429025	430336	431649	432964	43428
660	435600	436921	438244	439569	440896	442225	443556	444889	446224	44756
670	448900	450241	451584	452929	454276	455625	456976	458329	459684	46104
680	462400	463761	465124	466489	467856	469225	470596	471969	473344	47472
690	476100	477481	478864	480249	481636	483025	484416	485809	487204	48860
700	490000	491401	492804	494209	495616	497025	498436	499849	501264	50268
710	504100	505521	506944	508369	509796	511225	512656	514089	515524	51696
720	518400	519841	521284	522729	524176	525625	527076	528529	529984	53144

quares of integers, 1–999

	0	1	2	3	4	5	6	7	8	9
30	532900	534361	535824	537289	538756	540225	541696	543169	544644	546121
40	547600	549081	550564	552049	553536	555025	556516	558009	559504	561001
50	562500	564001	565504	567009	568516	570025	571536	573049	574564	576081
60	577600	579121	580644	582169	583696	585225	586756	588289	589824	591361
70	592900	594441	595984	597529	599076	600625	602176	603729	605284	606841
80	608400	609961	611524	613089	614656	616225	617796	619369	620944	622521
90	624100	625681	627264	628849	630436	632025	633616	635209	636804	638401
00	640000	641601	643204	644809	646416	648025	649636	651249	652864	654481
10	656100	657721	659344	660969	662596	664225	665856	667489	669124	670761
20	672400	674041	675684	677329	678976	680625	682276	683929	685584	687241
30	688900	690561	692224	693889	695556	697225	698896	700569	702244	703921
40	705600	707281	708964	710649	712336	714025	715716	717409	719104	720801
50	722500	724201	725904	727609	729316	731025	732736	734449	736164	737881
60	739600	741321	743044	744769	746496	748225	749956	751689	753424	755161
70	756900	758641	760384	762129	763876	765625	767376	769129	770884	772641
80	774400	776161	777924	779689	781456	783225	784996	786769	788544	790321
90	792100	793881	795664	797449	799236	801025	802816	804609	806404	808201
00	810000	811801	813604	815409	817216	819025	820836	822649	824464	826281
10	828100	829921	831744	833569	835396	837225	839056	840889	842724	844561
20	846400	848241	850084	851929	853776	855625	857476	859329	861184	863041
0	864900	866761	868624	870489	872356	874225	876096	877969	879884	881721
0	883600	885481	887364	889249	891136	893025	894916	896809	898704	900601
0	902500	904401	906304	908209	910116	912025	913936	915849	917764	919681
0	921600	923521	925444	927369	929296	931225	933156	935089	937024	938961
0	940900	942841	944784	946729	948676	950625	952576	954529	956484	958441
0	960400	962361	964324	966289	968256	970225	972196	974169	976144	978121
0	980100	982081	984064	986049	988036	990025	992016	994009	996004	998001

able 73
quare roots of integers, 1–99

	0	1	2	3	4	5	6	7	8	9
0	1·0000	1·4142	1·7321	2·0000	2·2361	2·4495	2·6458	2·8284	3·0000
	3·1623	3·3166	3·4641	3·6056	3·7417	3·8730	4·0000	4·1231	4·2426	4·3589
	4·4721	4·5826	4·6904	4·7958	4·8990	5·0000	5·0990	5·1962	5·2915	5·3852
	5·4772	5·5678	5·6569	5·7446	5·8310	5·9161	6·0000	6·0828	6·1644	6·2450
	6·3246	6·4031	6·4807	6·5574	6·6332	6·7082	6·7823	6·8557	6·9282	7·0000
	7·0711	7·1414	7·2111	7·2801	7·3485	7·4162	7·4833	7·5498	7·6158	7·6811
	7·7460	7·8102	7·8740	7·9373	8·0000	8·0623	8·1240	8·1854	8·2462	8·3066
	8·3666	8·4261	8·4853	8·5440	8·6023	8·6603	8·7178	8·7750	8·8318	8·8882
	8·9443	9·0000	9·0554	9·1104	9·1652	9·2195	9·2736	9·3274	9·3808	9·4340
	9·4868	9·5394	9·5917	9·6437	9·6954	9·7468	9·7980	9·8489	9·8995	9·9499

Table 74

Distribution of *t*

Degrees of freedom	Probability, p				
	0·1	0·05	0·02	0·01	0·001
1	6·31	12·71	31·82	63·66	636·62
2	2·92	4·30	6·97	9·93	31·60
3	2·35	3·18	4·54	5·84	12·92
4	2·13	2·78	3·75	4·60	8·61
5	2·02	2·57	3·37	4·03	6·87
6	1·94	2·45	3·14	3·71	5·96
7	1·89	2·37	3·00	3·50	5·41
8	1·86	2·31	2·90	3·36	5·04
9	1·83	2·26	2·82	3·25	4·78
10	1·81	2·23	2·76	3·17	4·59
11	1·80	2·20	2·72	3·11	4·44
12	1·78	2·18	2·68	3·06	4·32
13	1·77	2·16	2·65	3·01	4·22
14	1·76	2·14	2·62	2·98	4·14
15	1·75	2·13	2·60	2·95	4·07
16	1·75	2·12	2·58	2·92	4·02
17	1·74	2·11	2·57	2·90	3·97
18	1·73	2·10	2·55	2·88	3·92
19	1·73	2·09	2·54	2·86	3·88
20	1·72	2·09	2·53	2·85	3·85
21	1·72	2·08	2·52	2·83	3·82
22	1·72	2·07	2·51	2·82	3·79
23	1·71	2·07	2·50	2·81	3·77
24	1·71	2·06	2·49	2·80	3·75
25	1·71	2·06	2·49	2·79	3·73
26	1·71	2·06	2·48	2·78	3·71
27	1·70	2·05	2·47	2·77	3·69
28	1·70	2·05	2·47	2·76	3·67
29	1·70	2·05	2·46	2·76	3·66
30	1·70	2·04	2·46	2·75	3·65
40	1·68	2·02	2·42	2·70	3·55
60	1·67	2·00	2·39	2·66	3·46
120	1·66	1·98	2·36	2·62	3·37
∞	1·65	1·96	2·33	2·58	3·29

Table 74 is abridged from Table III of Fisher and Yates: *Statistical Tables for Biological, Agricultural and Medical Research*, Oliver & Boyd Ltd, Edinburgh, by permission of the authors and publishers.

Table 75

Variance ratio, F, for $p = 0.20$

Degrees of freedom, n_2	Degrees of freedom,								
	1	2	3	4	5	6	8	12	24	∞
1	9·5	12·0	13·1	13·6	14·0	14·3	14·6	14·9	15·2	15·6
2	3·6	4·0	4·2	4·2	4·3	4·3	4·4	4·4	4·4	4·5
3	2·7	2·9	2·9	3·0	3·0	3·0	3·0	3·0	3·0	3·0
4	2·4	2·5	2·5	2·5	2·5	2·5	2·5	2·5	2·4	2·4
5	2·2	2·3	2·3	2·2	2·2	2·2	2·2	2·2	2·2	2·1
6	2·1	2·1	2·1	2·1	2·1	2·1	2·0	2·0	2·0	2·0
7	2·0	2·0	2·0	2·0	2·0	2·0	1·9	1·9	1·9	1·8
8	2·0	2·0	2·0	1·9	1·9	1·9	1·9	1·8	1·8	1·7
9	1·9	1·9	1·9	1·9	1·9	1·8	1·8	1·8	1·7	1·7
10	1·9	1·9	1·9	1·8	1·8	1·8	1·8	1·7	1·7	1·6
11	1·9	1·9	1·8	1·8	1·8	1·8	1·7	1·7	1·6	1·6
12	1·8	1·8	1·8	1·8	1·7	1·7	1·7	1·7	1·6	1·5
13	1·8	1·8	1·8	1·8	1·7	1·7	1·7	1·6	1·6	1·5
14	1·8	1·8	1·8	1·7	1·7	1·7	1·6	1·6	1·6	1·5
15	1·8	1·8	1·8	1·7	1·7	1·7	1·6	1·6	1·5	1·5
16	1·8	1·8	1·7	1·7	1·7	1·6	1·6	1·6	1·5	1·4
17	1·8	1·8	1·7	1·7	1·7	1·6	1·6	1·6	1·5	1·4
18	1·8	1·8	1·7	1·7	1·6	1·6	1·6	1·5	1·5	1·4
19	1·8	1·8	1·7	1·7	1·6	1·6	1·6	1·5	1·5	1·4
20	1·8	1·8	1·7	1·7	1·6	1·6	1·6	1·5	1·5	1·4
21	1·8	1·7	1·7	1·7	1·6	1·6	1·6	1·5	1·4	1·4
22	1·8	1·7	1·7	1·6	1·6	1·6	1·5	1·5	1·4	1·4
23	1·7	1·7	1·7	1·6	1·6	1·6	1·5	1·5	1·4	1·3
24	1·7	1·7	1·7	1·6	1·6	1·6	1·5	1·5	1·4	1·3
25	1·7	1·7	1·7	1·6	1·6	1·6	1·5	1·5	1·4	1·3
26	1·7	1·7	1·7	1·6	1·6	1·6	1·5	1·5	1·4	1·3
27	1·7	1·7	1·7	1·6	1·6	1·6	1·5	1·5	1·4	1·3
28	1·7	1·7	1·7	1·6	1·6	1·6	1·5	1·5	1·4	1·3
29	1·7	1·7	1·7	1·6	1·6	1·5	1·5	1·5	1·4	1·3
30	1·7	1·7	1·6	1·6	1·6	1·5	1·5	1·5	1·4	1·3
40	1·7	1·7	1·6	1·6	1·5	1·5	1·5	1·4	1·3	1·2
60	1·7	1·7	1·6	1·6	1·5	1·5	1·4	1·4	1·3	1·2
120	1·7	1·6	1·6	1·5	1·5	1·5	1·4	1·4	1·3	1·1
∞	1·6	1·6	1·6	1·5	1·5	1·4	1·4	1·3	1·2	1·0

Table 75 is abridged from Table V of Fisher and Yates: *Statistical Tables for Biological, Agricultural and Medical Research*, Oliver & Boyd Ltd, Edinburgh. by permission of the authors and publishers.

Table 76

Variance ratio, F, for $p = 0.05$

Degrees of freedom, n_2 *Degrees of freedom,* n_1

n_2	1	2	3	4	5	6	8	12	24	∞
1	161·4	199·5	215·7	224·6	230·2	234·0	238·9	243·9	249·0	254·3
2	18·5	19·0	19·2	19·3	19·3	19·3	19·4	19·4	19·5	19·5
3	10·1	9·6	9·3	9·1	9·0	8·9	8·8	8·7	8·6	8·5
4	7·7	6·9	6·6	6·4	6·3	6·2	6·0	5·9	5·8	5·6
5	6·6	5·8	5·4	5·2	5·1	5·0	4·8	4·7	4·5	4·4
6	6·0	5·1	4·8	4·5	4·4	4·3	4·2	4·0	3·8	3·7
7	5·6	4·7	4·4	4·1	4·0	3·9	3·7	3·6	3·4	3·2
8	5·3	4·5	4·1	3·8	3·7	3·6	3·4	3·3	3·1	2·9
9	5·1	4·3	3·9	3·6	3·5	3·4	3·2	3·1	2·9	2·7
10	5·0	4·1	3·7	3·5	3·3	3·2	3·1	2·9	2·7	2·5
11	4·8	4·0	3·6	3·4	3·2	3·1	3·0	2·8	2·6	2·4
12	4·8	3·9	3·5	3·3	3·1	3·0	2·9	2·7	2·5	2·3
13	4·7	3·8	3·4	3·2	3·0	2·9	2·8	2·6	2·5	2·2
14	4·6	3·7	3·3	3·1	3·0	2·9	2·7	2·5	2·4	2·1
15	4·5	3·7	3·3	3·1	2·9	2·8	2·6	2·5	2·3	2·1
16	4·5	3·6	3·2	3·0	2·9	2·7	2·6	2·4	2·2	2·0
17	4·5	3·6	3·2	3·0	2·8	2·7	2·6	2·4	2·2	2·0
18	4·4	3·6	3·2	2·9	2·8	2·7	2·5	2·3	2·2	1·9
19	4·4	3·5	3·1	2·9	2·7	2·6	2·5	2·3	2·1	1·9
20	4·4	3·5	3·1	2·9	2·7	2·6	2·5	2·3	2·1	1·8
21	4·3	3·5	3·1	2·8	2·7	2·6	2·4	2·3	2·1	1·8
22	4·3	3·4	3·1	2·8	2·7	2·6	2·4	2·2	2·0	1·8
23	4·3	3·4	3·0	2·8	2·6	2·5	2·4	2·2	2·0	1·8
24	4·3	3·4	3·0	2·8	2·6	2·5	2·4	2·2	2·0	1·7
25	4·2	3·4	3·0	2·8	2·6	2·5	2·3	2·2	2·0	1·7
26	4·2	3·4	3·0	2·7	2·6	2·5	2·3	2·2	2·0	1·7
27	4·2	3·4	3·0	2·7	2·6	2·5	2·3	2·1	1·9	1·7
28	4·2	3·3	3·0	2·7	2·6	2·4	2·3	2·1	1·9	1·7
29	4·2	3·3	2·9	2·7	2·5	2·4	2·3	2·1	1·9	1·6
30	4·2	3·3	2·9	2·7	2·5	2·4	2·3	2·1	1·9	1·6
40	4·1	3·2	2·8	2·6	2·5	2·3	2·2	2·0	1·8	1·5
60	4·0	3·2	2·8	2·5	2·4	2·3	2·1	1·9	1·7	1·4
120	3·9	3·1	2·7	2·5	2·3	2·2	2·0	1·8	1·6	1·3
∞	3·8	3·0	2·6	2·4	2·2	2·1	1·9	1·8	1·5	1·0

Table 76 is abridged from Table V of Fisher and Yates: *Statistical Tables for Biological, Agricultural and Medical Research*, Oliver & Boyd Ltd, Edinburgh, by permission of the authors and publishers.

Table 77
Variance ratio, F, for $p = 0.01$

Degrees of freedom, n_2	Degrees of freedom, n_1									
	1	2	3	4	5	6	8	12	24	∞
1	4052	4999	5403	5625	5764	5859	5982	6106	6234	6366
2	98·5	99·0	99·2	99·3	99·3	99·3	99·4	99·4	99·5	99·5
3	34·1	30·8	29·5	28·7	28·2	27·9	27·5	27·1	26·6	26·1
4	21·2	18·0	16·7	16·0	15·5	15·2	14·8	14·4	13·9	13·5
5	16·3	13·3	12·1	11·4	11·0	10·7	10·3	9·9	9·5	9·0
6	13·7	10·9	9·8	9·2	8·8	8·5	8·1	7·7	7·3	6·9
7	12·3	9·6	8·5	7·9	7·5	7·2	6·8	6·5	6·1	5·7
8	11·3	8·7	7·6	7·0	6·6	6·4	6·0	5·7	5·3	4·9
9	10·6	8·0	7·0	6·4	6·1	5·8	5·5	5·1	4·7	4·3
10	10·0	7·6	6·6	6·0	5·6	5·4	5·1	4·7	4·3	3·9
11	9·7	7·2	6·2	5·7	5·3	5·1	4·7	4·4	4·0	3·6
12	9·3	6·9	6·0	5·4	5·1	4·8	4·5	4·2	3·8	3·4
13	9·1	6·7	5·7	5·2	4·9	4·6	4·3	4·0	3·6	3·2
14	8·9	6·5	5·6	5·0	4·7	4·5	4·1	3·8	3·4	3·0
15	8·7	6·4	5·4	4·9	4·6	4·3	4·0	3·7	3·3	2·9
16	8·5	6·2	5·3	4·8	4·4	4·2	3·9	3·6	3·2	2·8
17	8·4	6·1	5·2	4·7	4·3	4·1	3·8	3·5	3·1	2·7
18	8·3	6·0	5·1	4·6	4·3	4·0	3·7	3·4	3·0	2·6
19	8·2	5·9	5·0	4·5	4·2	3·9	3·6	3·3	2·9	2·5
20	8·1	5·9	4·9	4·4	4·1	3·9	3·6	3·2	2·9	2·4
21	8·0	5·8	4·9	4·4	4·0	3·8	3·5	3·2	2·8	2·4
22	7·9	5·7	4·8	4·3	4·0	3·8	3·5	3·1	2·8	2·3
23	7·9	5·7	4·8	4·3	3·9	3·7	3·4	3·1	2·7	2·3
24	7·8	5·6	4·7	4·2	3·9	3·7	3·4	3·0	2·7	2·2
25	7·8	5·6	4·7	4·2	3·9	3·6	3·3	3·0	2·6	2·2
26	7·7	5·5	4·6	4·1	3·8	3·6	3·3	3·0	2·6	2·1
27	7·7	5·5	4·6	4·1	3·8	3·6	3·3	2·9	2·6	2·1
28	7·6	5·5	4·6	4·1	3·8	3·5	3·2	2·9	2·5	2·1
29	7·6	5·4	4·5	4·0	3·7	3·5	3·2	2·9	2·5	2·0
30	7·6	5·4	4·5	4·0	3·7	3·5	3·2	2·8	2·5	2·0
40	7·3	5·2	4·3	3·8	3·5	3·3	3·0	2·7	2·3	1·8
60	7·1	5·0	4·1	3·7	3·3	3·1	2·8	2·5	2·1	1·6
120	6·9	4·8	4·0	3·5	3·2	3·0	2·7	2·3	2·0	1·4
∞	6·6	4·6	3·8	3·3	3·0	2·8	2·5	2·2	1·8	1·0

Table 77 is abridged from Table V of Fisher and Yates: *Statistical Tables for Biological, Agricultural and Medical Research*, Oliver & Boyd Ltd, Edinburgh, by permission of the authors and publishers.

Table 78
Variance ratio, F for $p = 0.001$

Degrees of freedom, n_2	Degrees of freedom, n_1									
	1	2	3	4	5	6	8	12	24	∞
1*	405	500	540	563	576	586	598	611	632	637
2	998·5	999·0	999·2	999·2	999·3	999·3	999·4	999·4	999·5	999·5
3	167·0	148·5	141·1	137·1	134·6	132·8	130·6	128·3	125·9	123·5
4	74·1	61·3	56·2	53·4	51·7	50·5	49·0	47·4	45·8	44·1
5	47·2	37·1	33·2	31·1	29·8	28·8	27·6	26·4	25·1	23·8
6	35·5	27·0	23·7	21·9	20·8	20·0	19·0	18·0	16·9	15·8
7	29·3	21·7	18·8	17·2	16·2	15·5	14·6	13·7	12·7	11·7
8	25·4	18·5	15·8	14·4	13·5	12·9	12·0	11·2	10·3	9·3
9	22·9	16·4	13·9	12·6	11·7	11·1	10·4	9·6	8·7	7·8
10	21·0	14·9	12·6	11·3	10·5	9·9	9·2	8·5	7·6	6·8
11	19·7	13·8	11·6	10·4	9·6	9·1	8·4	7·6	6·9	6·0
12	18·6	13·0	10·8	9·6	8·9	8·4	7·7	7·0	6·3	5·4
13	17·8	12·3	10·2	9·1	8·4	7·9	7·2	6·5	5·8	5·0
14	17·1	11·8	9·7	8·6	7·9	7·4	6·8	6·1	5·4	4·6
15	16·6	11·3	9·3	8·3	7·6	7·1	6·5	5·8	5·1	4·3
16	16·1	11·0	9·0	7·9	7·3	6·8	6·2	5·6	4·9	4·1
17	15·7	10·7	8·7	7·7	7·0	6·6	6·0	5·3	4·6	3·9
18	15·4	10·4	8·5	7·5	6·8	6·4	5·8	5·1	4·5	3·7
19	15·1	10·2	8·3	7·3	6·6	6·2	5·6	5·0	4·3	3·5
20	14·8	10·0	8·1	7·1	6·5	6·0	5·4	4·8	4·2	3·4
21	14·6	9·8	7·9	7·0	6·3	5·9	5·3	4·7	4·0	3·3
22	14·4	9·6	7·8	6·8	6·2	5·8	5·2	4·6	3·9	3·2
23	14·2	9·5	7·7	6·7	6·1	5·7	5·1	4·5	3·8	3·1
24	14·0	9·3	7·6	6·6	6·0	5·6	5·0	4·4	3·7	3·0
25	13·9	9·2	7·5	6·5	5·9	5·5	4·9	4·3	3·7	2·9
26	13·7	9·1	7·4	6·4	5·8	5·4	4·8	4·2	3·6	2·8
27	13·6	9·0	7·3	6·3	5·7	5·3	4·8	4·2	3·5	2·8
28	13·5	8·9	7·2	6·3	5·7	5·2	4·7	4·1	3·5	2·7
29	13·4	8·9	7·1	6·2	5·6	5·2	4·6	4·1	3·4	2·6
30	13·3	8·8	7·1	6·1	5·5	5·1	4·6	4·0	3·4	2·6
40	12·6	8·3	6·6	5·7	5·1	4·7	4·2	3·6	3·0	2·2
60	12·0	7·8	6·2	5·3	4·8	4·4	3·9	3·3	2·7	1·9
120	11·4	7·3	5·8	5·0	4·4	4·0	3·6	3·0	2·4	1·5
∞	10·8	6·9	5·4	4·6	4·1	3·7	3·3	2·7	2·1	1·0

* entries in this row must be multiplied by 1 000.

Table 78 is abridged from Table V of Fisher and Yates: *Statistical Tables for Biological, Agricultural and Medical Research*, Oliver & Boyd Ltd, Edinburgh, by permission of the authors and publishers.

Table 79

The correlation coefficient, r

Degrees of freedom	Probability, p				
	0·1	0·05	0·02	0·01	0·001
1	0·988	0·997	1·000	1·000	1·000
2	0·900	0·950	0·980	0·990	0·999
3	0·805	0·878	0·934	0·959	0·991
4	0·729	0·811	0·882	0·917	0·974
5	0·669	0·755	0·833	0·875	0·951
6	0·622	0·707	0·789	0·834	0·925
7	0·582	0·666	0·750	0·798	0·898
8	0·549	0·632	0·716	0·765	0·872
9	0·521	0·602	0·685	0·735	0·847
10	0·497	0·576	0·658	0·708	0·823
11	0·476	0·553	0·634	0·684	0·801
12	0·458	0·532	0·612	0·661	0·780
13	0·441	0·514	0·592	0·641	0·760
14	0·426	0·497	0·574	0·623	0·742
15	0·412	0·482	0·558	0·606	0·725
16	0·400	0·468	0·543	0·590	0·708
17	0·389	0·456	0·529	0·575	0·693
18	0·378	0·444	0·516	0·561	0·679
19	0·369	0·433	0·503	0·549	0·665
20	0·360	0·423	0·492	0·537	0·652
25	0·323	0·381	0·445	0·487	0·597
30	0·296	0·349	0·409	0·449	0·554
35	0·275	0·325	0·381	0·418	0·519
40	0·257	0·304	0·358	0·393	0·490
45	0·243	0·288	0·338	0·372	0·465
50	0·231	0·273	0·322	0·354	0·443
60	0·211	0·250	0·295	0·325	0·408
70	0·195	0·232	0·274	0·302	0·380
80	0·183	0·217	0·257	0·283	0·357
90	0·173	0·205	0·242	0·267	0·338
100	0·164	0·195	0·230	0·254	0·321

Table 79 is abridged from Table VII of Fisher and Yates: *Statistical Tables for Biological, Agricultural and Medical Research*, Oliver & Boyd Ltd, Edinburgh, by permission of the authors and publishers.

Table 80

Distribution of χ^2

Degrees of freedom					Probability, p							
	0·99	0·98	0·95	0·90	0·80	0·50	0·20	0·10	0·05	0·02	0·01	0·0
1	0·000	0·001	0·004	0·016	0·064	0·455	1·64	2·71	3·84	5·41	6·64	10·
2	0·020	0·040	0·103	0·211	0·446	1·386	3·22	4·61	5·99	7·82	9·21	13·
3	0·115	0·185	0·352	0·584	1·005	2·366	4·64	6·25	7·82	9·84	11·35	16·
4	0·297	0·429	0·711	1·064	1·649	3·357	5·99	7·78	9·49	11·67	13·28	18·
5	0·554	0·752	1·145	1·610	2·343	4·351	7·29	9·24	11·07	13·39	15·09	20·
6	0·872	1·134	1·635	2·204	3·070	5·35	8·56	10·65	12·59	15·03	16·81	22·
7	1·239	1·564	2·167	2·833	3·822	6·35	9·80	12·02	14·07	16·62	18·48	24·
8	1·646	2·032	2·733	3·490	4·594	7·34	11·03	13·36	15·51	18·17	20·09	26·
9	2·088	2·532	3·325	4·168	5·380	8·34	12·24	14·68	16·92	19·68	21·67	27·
10	2·558	3·059	3·940	4·865	6·179	9·34	13·44	15·99	18·31	21·16	23·21	29·
11	3·05	3·61	4·58	5·58	6·99	10·34	14·63	17·28	19·68	22·62	24·73	31·
12	3·57	4·18	5·23	6·30	7·81	11·34	15·81	18·55	21·03	24·05	26·22	32·
13	4·11	4·77	5·89	7·04	8·63	12·34	16·99	19·81	22·36	25·47	27·69	34·
14	4·66	5·37	6·57	7·79	9·47	13·34	18·15	21·06	23·69	26·87	29·14	36·
15	5·23	5·99	7·26	8·55	10·31	14·34	19·31	22·31	25·00	28·26	30·58	37·
16	5·81	6·61	7·96	9·31	11·15	15·34	20·47	23·54	26·30	29·63	32·00	39·
17	6·41	7·26	8·67	10·09	12·00	16·34	21·62	24·77	27·59	31·00	33·41	40·
18	7·02	7·91	9·39	10·87	12·86	17·34	22·76	25·99	28·87	32·35	34·81	42·
19	7·63	8·57	10·12	11·65	13·72	18·34	23·90	27·20	30·14	33·69	36·19	43·
20	8·26	9·24	10·85	12·44	14·58	19·34	25·04	28·41	31·41	35·02	37·57	45·
21	8·90	9·92	11·59	13·24	15·45	20·34	26·17	29·62	32·67	36·34	38·93	46·
22	9·54	10·60	12·34	14·04	16·31	21·34	27·30	30·81	33·92	37·66	40·29	48·
23	10·20	11·29	13·09	14·85	17·19	22·34	28·43	32·01	35·17	38·97	41·64	49·
24	10·86	11·99	13·85	15·66	18·06	23·34	29·55	33·20	36·42	40·27	42·98	51·
25	11·52	12·70	14·61	16·47	18·94	24·34	30·68	34·38	37·65	41·57	44·31	52·
26	12·20	13·41	15·38	17·29	19·82	25·34	31·80	35·56	38·89	42·86	45·64	54·
27	12·88	14·13	16·15	18·11	20·70	26·34	32·91	36·74	40·11	44·14	46·96	55·
28	13·57	14·85	16·93	18·94	21·59	27·34	34·03	37·92	41·34	45·42	48·28	56·
29	14·26	15·57	17·71	19·77	22·48	28·34	35·14	39·09	42·56	46·69	49·59	58·
30	14·95	16·31	18·49	20·60	23·36	29·34	36·25	40·26	43·77	47·96	50·89	59·

Table 80 is abridged from Table IV of Fisher and Yates: *Statistical Tables for Biologica Agricultural and Medical Research*, Oliver & Boyd Ltd, Edinburgh, by permission of th authors and publishers.

Table 81

The arcsin transformation

Gives the value of θ, in degrees, for which $\sin \theta = \sqrt{\left(\dfrac{\text{percentage}}{100}\right)}$

Percentage	0	1	2	3	4	5	6	7	8	9
0	0	5·7	8·1	10·0	11·5	12·9	14·2	15·3	16·4	17·5
10	18·4	19·4	20·3	21·1	22·0	22·8	23·6	24·4	25·1	25·8
20	26·6	27·3	28·0	28·7	29·3	30·0	30·7	31·3	32·0	32·6
30	33·2	33·8	34·4	35·1	35·7	36·3	36·9	37·5	38·1	38·6
40	39·2	39·8	40·4	41·0	41·5	42·1	42·7	43·3	43·8	44·4
50	45·0	45·6	46·1	46·7	47·3	47·9	48·5	49·0	49·6	50·2
60	50·8	51·3	51·9	52·5	53·1	53·7	54·3	54·9	55·6	56·2
70	56·8	57·4	58·0	58·7	59·3	60·0	60·7	61·3	62·0	62·7
80	63·4	64·2	64·9	65·7	66·4	67·2	68·0	68·9	69·7	70·6
90	71·6	72·5	73·6	74·7	75·8	77·1	78·5	80·0	81·9	84·3
100	90·0									

Proportions may be transformed by first multiplying them by 100.

Table 82

Factorial numbers

n	$n!$	n	$n!$
0	1	10	3 628 800
1	1	11	39 916 800
2	2	12	479 001 600
3	6	13	6 227 020 800
4	24	14	87 178 291 200
5	120	15	1 307 674 368 000
6	720	16	20 922 789 888 000
7	5 040		
8	40 320		
9	362 880		

Table 83
Numbers of tables in distribution-free tests

Number of observations n	Number of categories or treatments			
	2	3	4	5
6	**20**		**45**	
	1	—	2	—
	(1) 50000		(1) 22222	
7	**35**	**105**	**105**	**105**
	1	5	5	5
	(1) 28571	(2) 95238.	(2) 95238	(2) 95238
8	**70**	**280**	**105**	**420**
	3	14	5	21
	(1) 14286	(2) 35714	(2) 95238	(2) 23810
9	**126**	**280**	**1260**	**945**
	6	14	63	47
	(2) 79365	(2) 35714	(3) 79365	(2) 10582
10	**252**	**2100**	**6300**	**945**
	12	105	315	47
	(2) 39683	(3) 47619	(3) 15873	(2) 10582
11	**462**	**5775**	**15400**	**17325**
	23	288	770	866
	(2) 21645	(3) 17316	(4) 64935	(4) 57720
12	**942**	**5775**	**15400**	**138600**
	47	288	770	6930
	(2) 10823	(3) 17316	(4) 64935	(5) 72150
13	**1716**	**45045**	**200200**	**600600**
	85	2252	10010	30030
	(3) 58275	(4) 22200	(5) 49950	(5) 16650
14	**3432**	**126126**	**1051050**	**1401400**
	171	6306	52552	70070
	(3) 29138	(5) 79286	(6) 95143	(6) 71357
15	**6435**	**126126**	**2627625**	**1401400**
	321	6306	131381	70070
	(3) 15540	(5) 79286	(6) 38057	(6) 71357
16	**12870**	**1009008**	**2627625**	**28028000**
	643	50450	131381	140140
	(4) 77700	(6) 99107	(6) 38057	(7) 35679

This gives the numbers of distinguishable tables (bold figures) when the observations are distributed as evenly as possible between the various treatments, categories or columns. For example, if there are three treatments and 13 observations, the most even distribution is 4, 4 and 5, *any* two treatments consisting of four observations, and the other treatment consisting of five. If there were nine observations and two treatments, the most even distribution would be four to one treatment and five to the other. Beneath the numbers of tables in bold figures there are two entries which signify:

(1) number of tables × 0·05, rounded down.
(2) reciprocal of the number of tables, to five-figure accuracy. The number in brackets indicates the number of zeros following the decimal point.
    e.g. (4) 64935 is read as 0·000064935

To use these entries, see p. 133.

Table 84

Critical values of u

$n_1 = n_2$	$p \leqslant 0.001$		$p \leqslant 0.01$		$p \leqslant 0.05$	
4	—		—		2	0·029
5	—		2	0·008	3	0·040
6	—		2	0·002	3	0·013
7	2	0·0006	3	0·004	4	0·025
8	2	0·0002	4	0·009	5	0·032
9	3	0·0004	4	0·003	6	0·045
10	4	0·0010	5	0·004	6	0·018
11	4	0·0003	6	0·007	7	0·023
12	5	0·0005	7	0·009	8	0·030
13	5	0·0002	7	0·004	9	0·034
14	6	0·0004	8	0·006	10	0·041
15	7	0·0006	9	0·007	11	0·046
16	8	0·0009	10	0·009	11	0·023
17	8	0·0003	10	0·004	12	0·027
18	9	0·0005	11	0·005	13	0·030
19	10	0·0007	12	0·006	14	0·035
20	11	0·0009	13	0·008	15	0·038

Table 85

Critical values of T, for $p = 0{\cdot}05$

Number of observations, n_2	Number of observations, n_1													
	2	3	4	5	6	7	8	9	10	11	12	13	14	15
4			10											
5		6	11	17										
6		7	12	18	26									
7		7	13	20	27	36								
8	3	8	14	21	29	38	49							
9	3	8	15	22	31	40	51	63						
10	3	9	15	23	32	42	53	65	78					
11	4	9	16	24	34	44	55	68	81	96				
12	4	10	17	26	35	46	58	71	85	99	115			
13	4	10	18	27	37	48	60	73	88	103	119	137		
14	4	11	19	28	38	50	63	76	91	106	123	141	160	
15	4	11	20	29	40	52	65	79	94	110	127	145	164	185
16	4	12	21	31	42	54	67	82	97	114	131	150	169	
17	5	12	21	32	43	56	70	84	100	117	135	154		
18	5	13	22	33	45	58	72	87	103	121	139			
19	5	13	23	34	46	60	74	90	107	124				
20	5	14	24	35	48	62	77	93	110					
21	6	14	25	37	50	64	79	95						
22	6	15	26	38	51	66	82							
23	6	15	27	39	53	68								
24	6	16	28	40	55									
25	6	16	28	42										
26	7	17	29											
27	7	17												
28	7													

If the groups are unequal in size, n_1 refers to the smaller.

Table 85 is reproduced from White, C. (1952) *Biometrics* **8**: 33–41, by permission of the author and publishers.

Table 86

Critical values of T, for $p = 0.01$

Number of observations, n_2	2	3	4	5	6	7	8	9	10	11	12	13	14	15
5				15										
6			10	16	23									
7			10	17	24	32								
8			11	17	25	34	43							
9		6	11	18	26	35	45	56						
10		6	12	19	27	37	47	58	71					
11		6	12	20	28	38	49	61	74	87				
12		7	13	21	30	40	51	63	76	90	106			
13		7	14	22	31	41	53	65	79	93	109	125		
14		7	14	22	32	43	54	67	81	96	112	129	147	
15		8	15	23	33	44	56	70	84	99	115	133	151	171
16		8	15	24	34	46	58	72	86	102	119	137	155	
17		8	16	25	36	47	60	74	89	105	122	140		
18		8	16	26	37	49	62	76	92	108	125			
19	3	9	17	27	38	50	64	78	94	111				
20	3	9	18	28	39	52	66	81	97					
21	3	9	18	29	40	53	68	83						
22	3	10	19	29	42	55	70							
23	3	10	19	30	43	57								
24	3	10	20	31	44									
25	3	11	20	32										
26	3	11	21											
27	4	11												
28	4													

If the groups are unequal in size, n_1 refers to the smaller.

Table 86 is reproduced from White, C. (1952) *Biometrics* **8**: 33–41, by permission of the author and publishers.

Table 87
Critical values of r_s

Number of observations, n	$p = 0.05$	$p = 0.01$
4	1·000	—
5	0·900	1·000
6	0·829	0·943
7	0·714	0·893
8	0·643	0·833
9	0·600	0·783
10	0·564	0·746
12	0·506	0·712
14	0·456	0·645
16	0·425	0·601
18	0·399	0·564
20	0·377	0·534
22	0·359	0·508
24	0·343	0·485
26	0·329	0·465
28	0·317	0·448
30	0·306	0·432

Table 87 is reproduced from Siegel, S. (1956) *Non-parametric statistics*, McGraw-Hill Book Company, New York, by permission of the author and publishers.

Table 88
Significance levels for choice-chamber and similar experiments

Number of trials	$p \leqslant 0.01$		$p \leqslant 0.05$	
6	—		6 + 0	0·031
10	10 + 0	0·002	9 + 1	0·021
15	13 + 2	0·008	12 + 3	0·035
20	17 + 3	0·003	15 + 5	0·038
25	21 + 4	0·009	17 + 8	0·043
30	22 + 8	0·006	20 + 10	0·034

Table 88 is based on Bishop, O.N. (1971), *Outdoor Biology; Teachers' Guide*, John Murray Ltd., by permission of the publishers.

Bibliography

BROOKES, B. C., and DICK, W. F. L. (1951) *Introduction to Statistical Method*, Heinemann.

BROWNLEE, K. A. (1948) *Industrial Experimentation*, H.M.S.O.

COCHRAN, W. G. (1964) Approximate significance levels of the Behrens-Fisher Test: *Biometrics* **20**, 191–195. (Comparing two samples when one cannot assume variances are equal.)

COX, D. R. (1958) *Planning of experiments*, John Wiley.

EDGINGTON, E. S. (1969) *Statistical inference*, McGraw-Hill.

FISHER, Sir R. A. (1960) *The Design of Experiments*, 7th edn, Oliver & Boyd.

FISHER, Sir R. A. (1963) *Statistical Methods for Research Workers*, 13th edn, Oliver & Boyd.

FISHER, Sir R. A., and YATES, F. (1963) *Statistical Tables for Biological, Agricultural and Medical Research*, 6th edn, Oliver & Boyd.

GREIG-SMITH, P. (1964) *Quantitative Plant Ecology*, Butterworth.

HOLMAN, H. H. (1962) *Biological Research Method*, Oliver & Boyd.

KERSHAW, K. A. (1964) *Quantitative and Dynamic Ecology*, Edward Arnold.

LEWIS, T., and TAYLOR, L. R. (1967) *Introduction to Experimental Ecology*, Academic Press.

LINDLEY, D. V., and MILLER, J. C. P. (1962) *Cambridge Elementary Statistical Tables*, Cambridge University Press.

SIEGEL, S. (1956) *Non-parametric statistics*, McGraw-Hill.

SNEDECOR, G. W. (1956) *Statistical Methods*, 5th edn, Iowa State College Press.

YEOMANS, K. A. (1968) *Statistics for the social scientist*, 2 vols., Penguin Books.

Answers to problems

Chapter 2, p. 13

1 5 ray-florets.

2 Though both histograms show a peak at 15–19 mm, those with no fertiliser have more plants below the peak, while those with fertiliser have more plants above the peak. Since there is overlap, it is not possible to state definitely that fertiliser increases growth.

3 The most useful class interval is about 0·02 g.

4 The histograms show little overlap, so it appears likely that the flowers, on average, contain more stamens than carpels. The histograms differ in shape: that for stamens is wider and flatter, suggesting that the number of stamens has greater variability.

5 Respiratory movements appear to be distinctly faster at the higher temperature.

6 The most useful class interval is about 10 mm. The histogram confirms the visual impression.

Chapter 3, p. 19

This table gives the answers and, for checking calculations, values of n and Σx:

Problem		Modal class	Median	n	Σx	Mean
1		40–44 ray-florets	44 ray-florets	51	2261	44·33 ray-florets
2	No fertiliser	15–19 mm	16 mm	23	365	15·87 mm
	Fertiliser	15–19 mm	19 mm	34	682	20·06 mm
3		0·50–0·69 g	0·82 g	19	15·21	0·801 g
4	Stamens	50–59 stamens	54 stamens	51	2783	54·57 stamens
	Carpels	30–39 carpels	30 carpels	51	1502	29·45 carpels
5	13°C	30–34 beats	33·5 beats	10	350	35·0 beats
	20°C	45–49 beats	51·5 beats	10	528	52·8 beats
6	Small-flowered	11 mm*	11 mm	16	171	10·69 mm
	Normal-flowered	19 mm*	19 mm	46	846	18·39 mm

* Class interval = 1 mm.

7 Means: 1·33 daisy plants, 0·63 ribwort plants; *Totals:* 53 200 daisy plants, 25 200 ribwort plants.

Chapter 3, p. 27

This table gives the answers and, for checking calculations, values of Σx^2 and Σd^2. To save space, units of measurement are given only in the range column.

Problem	Range	Lower quartile point	Upper quartile point	Semi-inter-quartile range	Σx^2	Σd^2	Variance s^2	Standard dev s
	23 ray-florets	41	48	3·5	101573	1335	26·176	5·116
No fertiliser	20 mm	13	18	2·5	6153	361	15·696	3·962
Fertiliser	21 mm	17	24	3·5	14530	850	25·000	5·000
	0·87 mm	0·59	0·95	0·18	13·2293	1·0533	0·05544	0·2355
Stamens	43 stamens	46	62	8	158167	6303	123·59	11·117
Carpels	25 carpels	26	33	3·5	45822	1587	31·12	5·579
13°C	10 beats	33	37	2	12350	100	10·00	3·162
20°C	25 beats	45	59	7	28466	587·6	58·76	7·666
Small-flowered	8 mm	9	12	1·5	1891	63	3·938	1·98
Normal-flowered	13 mm	17	20	1·5	15944	385	8·370	2·89

Notes: 2. The two distributions have almost equal ranges, but that of the plants with no fertiliser has a distinctly smaller variance and standard deviation, as was suggested by the appearance of the histogram.

4. Distribution of stamens has the greater dispersion.

6. Though the means of these distributions differ markedly, their dispersions are more or less equal.

Chapter 4, p.33

To save space, units of measurement are given only in the mean (μ) column.

Problem	Population mean μ	Population variance σ^2	Population standard deviation, σ
1	44·33 ray-florets	26·700	5·167
2 No fertiliser	15·87 mm	16·409	4·051
Fertiliser	20·06 mm	25·758	5·075
3	0·801 g	0·05852	0·2419
4 Stamens	54·57 stamens	126·06	11·228
Carpels	29·45 carpels	31·74	5·634
5 13°C	35·0 beats	11·11	3·333
20°C	52·8 beats	65·29	8·080
6 Small-flowered	10·69 mm	4·200	2·049
Normal-flowered	18·39 mm	8·556	2·925

Chapter 4, p. 34

8 $p = \frac{1}{6} = 0.167$ or 16·7%.

9 For odd or even, $p = \frac{3}{6} = 0.5$ or 50%.

10 $p = \frac{4}{52} = 0.0769$ or 7·69%.

$$p = \frac{1}{52} = 0.0192 \text{ or } 1.92\%.$$

$$p = \frac{13}{52} = 0.25 \text{ or } 25\%.$$

Chapter 4, p. 37

Problem	When $p=0.05$ $d=1.96$	When $p=0.01$ $d=2.58$
1	34·20 and 54·46	31·00 and 57·66
2 No fertiliser	7·93 and 23·81	5·42 and 26·32
Fertiliser	10·11 and 30·01	6·97 and 33·15
3	0·327 and 1·275	0·177 and 1·425
4 Stamens	32·56 and 76·58	25·60 and 83·54
Carpels	18·41 and 40·49	14·91 and 43·99
5 13°C	28·5 and 41·5	26·4 and 43·6
20°C	37·0 and 68·6	32·0 and 73·6
6 Small-flowered	6·67 and 14·71	5·40 and 15·98
Normal-flowered	12·66 and 24·12	10·84 and 25·94

Chapter 4, p. 39

This table gives also the values of *t* used in calculating the correct limits:

Problem	When $p=0.05$ t	Limits	When $p=0.01$ t	Limits
1	2·01*33·94	and 54·72	2·68*30·48	and 58·18
2 No fertiliser	2·07 7·48	and 24·26	2·82 4·45	and 27·29
Fertiliser	2·04 9·71	and 30·41	2·75 6·10	and 34·02
3	2·10 0·293	and 1·309	2·88 0·104	and 1·498
4 Stamens	2·01*32·00	and 77·14	2·68*24·48	and 84·66
Carpels	2·01*18·13	and 40·77	2·68*14·35	and 44·55
5 13°C	2·26 27·5	and 42·5	3·25 24·2	and 45·8
20°C	2·26 34·5	and 71·1	3·25 26·5	and 79·1
6 Small-flowered	2·13 6·33	and 15·05	2·95 4·65	and 16·73
Large-flowered	2·02 12·48	and 24·30	2·69*10·52	and 26·26

* denotes values of *t* estimated by interpolation in the table.

Chapter 4, p. 45

1 44·33 ±1·45 ray-florets.

2 No fertiliser: 15·87 ±1·75 mm.   *Fertiliser:* 20·06 ±1·78 mm.

3 0·801 ±0·117 g.

4 Stamens: 54·57 ±3·16 stamens.   *Carpels:* 29·45 ±1·59 carpels.

5 13°C: 35·0 ±2·4 beats.   *20°C:* 52·8 ±5·8 beats.

6 Small-flowered: 10·69 ±1·09 mm.   *Normal-flowered:* 18·39 ±0·87 mm.

Chapter 5, p. 50

2 $\sigma_d = 1.2128$, $t = 3.45$, 55 degrees of freedom; From table, at $p = 0.01$, and

60 degrees freedom, $t = 2.66$. Effect of fertiliser treatment is significant, at 1% level.

4 $\sigma_d = 1.759$, $t = 14.28$, 100 degrees of freedom. From table, at $p = 0.001$, and 120 degrees of freedom, $t = 3.37$. Flowers of Ranunculus repens contain on average more stamens than carpels; statement has better than 0.1% significance.

5 $\sigma_d = 2.764$, $t = 6.44$, 18 degrees of freedom. From table, at $p = 0.001$, and 18 degrees of freedom, $t = 3.92$. Respiratory movements are more rapid at 20°C, than at 13°C; statement has better than 0.1% significance.

6 $\sigma_d = 0.6697$, $t = 11.50$, 60 degrees of freedom. From table, at $p = 0.001$, and 60 degrees of freedom, $t = 3.46$. The plants picked out by eye as being 'small-flowered' have been shown by measurement to belong to a group having distinctly smaller flowers than normal plants; statement has better than 0.1% significance.

11 *Sunny bed:* $\Sigma x^2 = 890.25$, $\Sigma d^2 = 5.80$, $\sigma = 1.2042$, $\sigma_n = 0.5386$. Mean = $13.3 \pm 1.5°C$.

Evergreen shade: $\Sigma x^2 = 490.25$, $\Sigma d^2 = 0.20$, $\sigma = 0.2236$, $\sigma_n = 0.1000$. Mean = $9.9 \pm 0.3°C$.

Both distributions: $\sigma_d = 0.5477$, $t = 6.21$, 8 degrees of freedom. From table, at $p = 0.001$, and 8 degrees of freedom, $t = 5.04$. Soil temperatures in the sunny bed are higher than those in evergreen shade; statement has 0.1% significance.

12 *No fertiliser:* Mean = 15.70 mm, $\Sigma d^2 = 361$, as before, so all derived values are as before, including limits of mean.

Fertiliser: $\Sigma d^2 = 960$, $\sigma = 5.394$, $\sigma_n = 0.9250$. Mean = 19.79 ± 1.89 mm.

Both distributions: $\sigma_d = 1.2526$, $t = 3.27$, 55 degrees of freedom; From table, at $p = 0.1$, and 60 degrees of freedom, $t = 2.66$. Effect of fertiliser treatment is significant, at 1% level. (Compare with Problem No. 2 above.)

13 *IAA:* $\Sigma d^2 = 18618$, $\sigma = 43.15$, $\sigma_n = 13.01$. Mean = 66.8 ± 29.0 mm.

Control: $\Sigma d^2 = 33455$, $\sigma = 57.84$, $\sigma_n = 17.44$. Mean = $95.9 + 38.9$ mm.

Both distributions: $\sigma_d = 21.76$, $t = 1.34$, 20 degrees of freedom. From table, at $p = 0.1$, and 20 degrees of freedom, $t = 1.72$. Difference between treatments is approaching 10% significance level; an indication that some effect may exist, but requires further experiments to obtain more data.

14 *Maternal age:* $\Sigma d^2 = 1221$, $\sigma = 8.016$, $\sigma_n = \sigma_d = 1.792$. Mean = 31.55 ± 3.75 years. Comparison with mean of general population gives $t = 1.96$, with infinite number of degrees of freedom. From table, at $p = 0.05$, $t = 1.96$. The mean maternal age is significantly different from mean maternal age of general population; statement has 5% significance (just).

Paternal age: $\Sigma d^2 = 1747$, $\sigma = 9.852$, $\sigma_n = \sigma_d = 2.260$. Mean = 34.95 ± 4.75 years. Comparison with mean of general population gives $t = 1.73$. with infinite number of degrees of freedom. From table, at $p = 0.10$, $t = 1.65$, and at $p = 0.05$, $t = 1.96$. The mean age is not significantly different from mean paternal age of general population at 5% level. Difference is significant at 10% level, an indication that further data might give more significant results.

Chapter 5, p. 55

4 Stamens: $\sigma_\sigma = 1{\cdot}1117$; taking $t = 2{\cdot}68$, 1 % limits are 8·250 and 14·208.

Carpels: $\sigma_\sigma = 0{\cdot}5578$; taking $t = 2{\cdot}68$, 1 % limits are 4·139 and 7·129. No overlap between stamen and carpel limits.

$F = 3{\cdot}97$, $n_1 = 50$ (nearest, ∞), $n_2 = 50$. Significant ratio when $p = 0{\cdot}002$ (double values of p from the tables).

5 $F = 5{\cdot}88$, $n_1 = 9$ (nearest, 8), $n_2 = 9$. Significant ratio when $p = 0{\cdot}02$.

6 $F = 2{\cdot}04$, $n_1 = 45$ (nearest, ∞), $n_2 = 15$. Exceeds tabulated F, when $p = 0{\cdot}40$, but this is not acceptable as a significant level of probability. Not significant.

11 $F = 29$, $n_1 = n_2 = 4$. Significant ratio, when $p = 0{\cdot}02$.

13 $F = 1{\cdot}65$, $n_1 = n_2 = 10$. Not significant.

14 $F = 1{\cdot}51$, $n_1 = 18$, $n_2 = 19$. Not significant.

Chapter 6, p. 62

15 $\Sigma x = 161{\cdot}5$; $A = 1795{\cdot}75$; $B = 1788{\cdot}55$; $D = 1738{\cdot}82$.

Source of variance	Sums of squares	Degrees of freedom	Mean squares
Between habitats	49·73	2	24·87
Residual	7·20	12	0·60
Total	56·93	14	

$F = 24{\cdot}87/0{\cdot}69 = 41{\cdot}5$, $n_1 = 2$, $n_2 = 12$. From Table 78, when $p = 0{\cdot}001$, $F = 13{\cdot}0$. Effect of habitat is significant at 0·1 % level.

16 $\Sigma x = 110$, $A = 1184{\cdot}0$, $B = 1126{\cdot}8$, $D = 806{\cdot}7$.

Source of variance	Sums of squares	Degrees of freedom	Mean squares
Between habitats	320·1	2	160·1
Residual	57·2	12	4·77
Total	377·3	14	

$F = 160{\cdot}1/4{\cdot}77 = 33{\cdot}6$, $n_1 = 2$, $n_2 = 12$. From Table 78, when $p = 0{\cdot}001$, $F = 13{\cdot}0$. Effect of habitat is significant at 0·1 % level.

17a $\Sigma x = 28{\cdot}37$; $A = 97{\cdot}1985$; $B = 93{\cdot}8600$; $D = 89{\cdot}4285$.

Source of variance	Sums of squares	Degrees of freedom	Mean squares
Between depths	4·4315	2	2·2158
Residual	3·3385	6	0·5564
Total	7·7700	8	

$F = 3{\cdot}98$, $n_1 = 2$, $n_2 = 8$. Significant only for $p = 0{\cdot}20$, not usually acceptable as a significant result.

17b Σx, A, B, D, as above; $C = 91{\cdot}8522$.

Source of variance	Sums of squares	Degrees of freedom	Mean squares
Between depths	4·4315	2	2·2158
Between sites	2·4237	2	1·2119
Residual	0·9148	4	0·2287
Total	7·7700	8	

For effect of depth: $F = 9.69$, $n_1 = 2$, $n_2 = 4$. Effect of depth significant for $p = 0.05$. Removal of between-sites sum of squares from residual has improved resolution of analysis, and effect of depth is now seen to be significant.

For effect of site: $F = 5.30$, $n_1 = 2$, $n_2 = 4$. Significant at $p = 0.20$, almost significant at $p = 0.05$. An indication of some effect, but since this was not of relevance to the investigation this effect was not studied further. The real reason for the two-factor analysis was to establish the significance of the effect of depth.

Calculation by machine

15 $G = 161.5$; $H = 1795.75$, $J = 8942.75$.
16 $G = 110$, $H = 1184$, $J = 5634.0$.
17 $G = 28.37$, $H = 97.1985$, $J = 281.5801$, $K = 275.5565$.

Chapter 7, p. 70

4 Data from 51 flowers: $\Sigma x = 2783$ (stamens), $\Sigma y = 1502$ (carpels), $\Sigma x^2 = 158167$, $\Sigma y^2 = 45822$, $\Sigma xy = 82155$, $\Sigma d_x^2 = 6303$, $\Sigma d_y^2 = 1587$, $\Sigma d_x d_y = 193$. $r = 0.061$, 50 degrees of freedom. Correlation not significant.

Data from first 20 flowers: $\Sigma x = 1099$ (stamens), $\Sigma y = 547$ (carpels), $\Sigma x^2 = 62319$, $\Sigma y^2 = 15527$, $\Sigma xy = 30041$, $\Sigma d_x^2 = 1929$, $\Sigma d_y^2 = 567$, $\Sigma d_x d_y = -17$. $r = -0.016$, 19 degrees of freedom. Correlation not significant.

18 $\Sigma x = 2879$ (mm), $\Sigma y = 458$ (spines), $\Sigma x^2 = 171725$, $\Sigma y^2 = 5782$, $\Sigma xy = 27872$, $\Sigma d_x^2 = 5952$, $\Sigma d_y^2 = 1587$, $\Sigma d_x d_y = 1500$. $r = 0.488$, 49 degrees of freedom. Correlation positive, and significant at 0.1% level. $x = 48.92 + 0.95y$. If $y = 6$, $x = 54.6$ mm. $y = -5.35 + 0.25x$. If $x = 70$, $y = 12$ spines.

19 $\Sigma x = 3150$ (μg/cm³), $\Sigma y = 17.7$ (meter reading), $\Sigma x^2 = 2047500$, $\Sigma y^2 = 78.11$, $\Sigma xy = 6105$, $\Sigma d_x^2 = 393750$, $\Sigma d_y^2 = 25.895$, $\Sigma d_x \Sigma d_y = -3187.5$. $r = -0.998$, 5 degrees of freedom. Correlation negative, and significant at 0.1% level, $x = 888 - 123y$. If $y = 1.3$, $x = 728$ μg/cm³.

Chapter 8, p. 78

20 $E = 13.75$ (white) and 41.25 (green); $\chi^2 = 0.297$; 1 degree of freedom; p is between 0.50 and 0.80; deviation not significant; results agree with theory.

21 $E = 42$ female and 42 male; $\chi^2 = 1.190$; 1 degree of freedom; p is between 0.20 and 0.50; deviation not significant; results agree with theory.

22 $E = 40$ (red), 80 (purple) and 40 (white); $\chi^2 = 4.400$; 2 degrees of freedom; p is between 0.10 and 0.20; deviation not significant; results agree with theory.

23 $E = 53.25$ (wild) and 17.75 (vestigial); $\chi^2 = 1.695$; 1 degree of freedom; p is between 0.10 and 0.20, almost 0.20; deviation not significant; results agree with theory.

24 $\chi^2 = 108$; 1 degree of freedom; p much less than 0.001; deviation significant; effect of breathing is significant; hypothesis is supported.

Statistics for biology

Chapter 10, p. 127

25 *1st area:* $\Sigma\theta = 109 \cdot 4°$, $\Sigma\theta^2 = 3996 \cdot 58$, $\Sigma d^2 = 7 \cdot 13$, $\sigma^2 = 3 \cdot 565$, $\theta = 36 \cdot 5° \equiv$ $\bar{x} = 35\%$.

2nd area: $\Sigma\theta = 178 \cdot 8°$, $\Sigma\theta^2 = 10726 \cdot 56$, $\Sigma d^2 = 70 \cdot 08$, $\sigma^2 = 35 \cdot 04$, $\theta = 59 \cdot 6°$ $\equiv \bar{x} = 74\%$.

Comparison: $\sigma_d = 3 \cdot 587$, $t = 6 \cdot 43$, 4 degrees of freedom. Significant difference of means at 1% level.

26 Use $\sqrt{(x + 0 \cdot 5)}$ transformation, denoted below by ΣX, ΣX^2, $\bar{X}$.

Shaded: $\Sigma X = 75 \cdot 7829$, $\Sigma X^2 = 133 \cdot 0$, $\Sigma d^2 = 18 \cdot 1$, $\sigma^2 = 0 \cdot 369$, $\bar{X} = 1 \cdot 51566$ $\equiv \bar{x} = 1 \cdot 80$.

Sunny: $\Sigma X = 46 \cdot 2158$, $\Sigma X^2 = 50 \cdot 00$, $\Sigma d^2 = 7 \cdot 28$, $\sigma^2 = 0 \cdot 149$, $\bar{X} = 0 \cdot 92432 \equiv$ $\bar{x} = 0 \cdot 35$.

Comparison: $\sigma_d = 0 \cdot 10178$, $t = 5 \cdot 81$, 98 degrees of freedom. Significant difference of means at 0.1% level, assuming that the data are normally distributed, which is questionable. In practice it would not be worth while to bother with a transformation; here it is done for the purposes of the exercise in technique of transformation.

27 $\chi^2 = 11 \cdot 3$, 1 degree of freedom. Negative association, significant at 0.1% level. With the small area of the ring, this is probably due to the presence of a plant or cluster of plants of one species tending to crowd out plants of the other species.

28 $\chi^2 = 0 \cdot 96$, 1 degree of freedom. p lies between 0.20 and 0.50. The numbers of even and odd do not differ significantly from 25.5 of each. No tendency to have an even number of spines is demonstrated.

Chapter 11, p. 131

29 16! ways = 20922789888000 ways.

30 9! ways = 362880 ways.

31 6!/4!2! ways = 15 ways.

32 1023!/5! 1018! ways = 9245818873599 ways.

Chapter 11, p. 145

33 14!/7!7! = 3432 possible tables; 1.48 g can be exchanged with 3 greater values under 'germinated seeds', giving 4 tables, including the observed data; each can give 4 by exchange of the 2 pairs of tied values, making a total of $4 \times 2 \times 4 = 16$ tables. $p = 16/3432 = 0 \cdot 0047$. Result significant.

34 $n_1 = 17$, $n_2 = 19$, $u = 10$. Read table 84 for $n_1 = n_2 = 18$; Result is significant for $p = 0 \cdot 005$.

35 $\mu_u = 76$; $\sigma_u^2 = 36 \cdot 21$; $t = 6 \cdot 5$. $p < 0 \cdot 0005$.

36 Tie for 15mm is resolved at random, and it makes no difference which way it is resolved. $n_1 = 10$, $n_2 = 14$, $u = 6$. Read table 84 for $n_1 = n_2 = 12$. Result is significant for $p = 0 \cdot 009$.

37 Ties are within samples, so cause no problems. $n_1 = n_2 = 7$, $u = 4$. Result is significant for $p = 0 \cdot 025$.

38 Median lies between 49 and 50. Sequence is bbbbbbbabaaaaaaa. $n_1 = n_2 = 8$, $u = 4$. Significant for $p = 0.009$.

39 $15!/5!5!5!3! = 126126$ possible tables; 9% (rank 11) can be interchanged with 8% (rank 10) to give greater sum of squares; no other tables to consider. $p = 2/126126 = 0.00016$. Highly significant result. With extra observation: $16!/5!6!5!2! = 1009008$ possible tables; 9% and 8% can be interchanged, and also the tied 4% values; $p = 4/1009008 = 0.000004$. Highly significant.

40 $9!/3!3!3!3! = 280$ possible tables. It is easy to write more than 14 tables with sums of squares equal to or greater than that of the observed table, so $p < 0.05$ and effect is not statistically significant.

41 For small-flowered buttercups, $T = 67$, $T' = 183$. For $n_1 = 10$, $n_2 = 14$ Table 86 gives critical value 81. Significant for $p = 0.01$.

42 For IAA treatment, $T = 104$, $T' = 149$. For $n_1 = n_2 = 11$, Table 85 gives critical value 96. Result not significant for $p = 0.05$.

43 $\Sigma d^2 = 2.5$, $r_s = 0.928$. Significant for $p = 0.05$.

Index

Bold figures refer to pages on which terms are defined or explained

215